Our
Human Ancestors

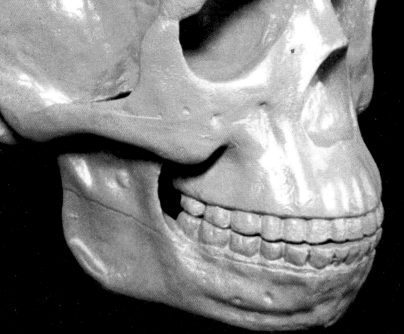

Our Human Ancestors

WARWICK PRESS, NEW YORK

Adviser

J.L. Angel, A.B., Ph.D.

Consultant Editor

Don Brothwell, B.Sc., M.A.

Editor

Frances M. Clapham

Assistant Editor

Abigail Frost B.A.

Contributors

Peter Andrews M.A., Ph.D.
Don Brothwell B.Sc., M.A.
Elisabeth Carter B.A.
Jonathan H. Musgrave M.A., Ph.D.

Robin Place M.A.
Nigel Seeley B.Sc., Ph.D. M.R.I.C., M.Inst.P.
Jane Siegel M.A.
C. Stringer B.Sc., Ph.D.

Illustrators

John Barber
Brian Dear
Constance Dear
Richard Eastland

Richard Hook
Peter Luff
Angus McBride
John Sibbick

Copyright © 1976 by Grisewood & Demsey Ltd.
Published in the U.S. 1977 by Warwick Press,
730 Fifth Avenue, New York, New York 10019
Printed in Great Britain by Purnell & Sons Ltd., Paulton
All rights reserved
6 5 4 3 2 1

Library of Congress Catalog Card No. 76—41664
ISBN 0–531–01279–4 lib. bdg.
ISBN 0–531–02481–4

Contents

Opposite, top left: Skull of a bushman; below left: terracotta figure from Mohenjo-Daro. Opposite right: Solutrean laurel-leaf blade. Below: A ceremonial knife from Peru. Right: A neolithic temple in Malta. Below right: A native of New Guinea. Title page: Set against a cave painting from Altamira are the figures of (from left to right) Plesiadapis, Aegyptopithecus, Ramapithecus, Homo erectus, Neanderthal man, Cro-Magnon man.

The Rise of Man

The story of man is one of survival through successful adaptation. The first part of this story traces his evolution from early primate ancestors to his present form; the second, the cultural evolution which enabled him to evolve complex technologies and sciences, to exploit his environment and to shape it to his own needs.

The early part of man's story lasted for millions of years, and much of it is still obscure. For evolution is a most complex process, and various species appeared and became extinct during the period before real man. Evidence provided by bones – often by only fragments of bones – has been meticulously studied in efforts to disentangle the complex evolutionary web, to establish which threads died out and which could lead to modern man. New finds may throw fresh light on long-held theories, leading to re-evaluations and occasionally reclassifications. Sometimes there may be only a single skull, or part of a skull, to link the evolutionary chain. But gradually the picture of man's development and of his spread through the world is emerging.

By about 40,000 years ago modern man – *Homo sapiens sapiens* – had fully evolved physically. He differed from all other animals in his combination of upright stance, relatively huge brain, stereoscopic colour vision, and manual dexterity. Even before this his ancestors had learned how to make and to use tools, and to harness fire; almost certainly they had developed language. He was a skilled hunter and food gatherer, usually leading a nomadic life dictated by the seasonal food supply.

The second part of man's story occupies, in comparison, only the briefest of time spans. At first *Homo sapiens sapiens* continued the nomadic, hunting and gathering life of his ancestors. Then came a change in the environment with the ending of the latest Ice Age, and the disappearance from many regions of the cold-climate animals on which he had preyed. But his adaptability enabled man to survive; he **first husbanded wild plants and animals, then cultivated and domes-**ticated them. For the first time he built permanent settlements, and cleared forested areas to grow his crops. Now he was shaping his environment to his own needs.

At first men were preoccupied with the overwhelming need for survival – enough to eat. But farmers were soon able to provide a food surplus to support non-producers such as priests and administrators. From now on populations soared; people became concentrated into villages, towns, and cities. Work became more and more specialized, and technological development was given a new impetus with the development of metallurgy. Complex systems of writing developed; and through trade men came into contact with other civilizations. From this point our human past is a matter of recorded history.

The study of man's rise to civilization is fascinating in itself. It may also be of value to us. For by giving us insight into the influences that have shaped his behaviour, it may help us to gain a better idea of how to survive successfully into the future.

The last stronghold of the Incas, Machu Picchu, was never discovered by the Spanish. The agricultural terraces which fell steeply away from the citadel, perched high among the Andes, enabled the Peruvians to survive. But the Spaniards had brought to an end a flourishing and sophisticated civilization that had grown up independently from those of the Old World.

9

What is Man?

Man, like all living creatures, has evolved from progressively simpler ancestors and ultimately from a unicellular creature 3000 million years ago. Survival is the main preoccupation of all living creatures, from the amoeba to man. Indeed until very recently most members of the human species devoted the major part of each day to keeping themselves and their species alive. Today food still has to be found, eaten, and digested to provide man with energy; waste products have to be excreted; and the members of the next generation have to be produced to carry on the species.

The parts of man's body that perform these vital functions are not markedly different from their counterparts in other mammals. So what are the features of man that set him apart from every other living creature, including the great apes? And, indeed, what do we mean by the term 'man'?

Dictionary compilers take no chances when they define 'man' as 'The human creature or race viewed as a genus of animals (*Homo*: in this definition consisting of only one species, *H. sapiens*)'. Philosophers have shown far more imagination and occasionally have anticipated the biological approach of modern anthropologists. They have variously described man as 'a two-legged animal . . . with broad, flat nails' (Plato); 'a political animal' (Aristotle); 'a social animal' (Carlyle); 'a tool-making animal' (Franklin). Anthropologists define man in much more formal terms. To them all mankind are members of the genus *Homo*; and the defining characters of the genus have been listed by one authority as follows:

> *Homo is a genus of the hominidae with the following characters: the structure of the pelvic girdle and the hindlimb skeleton is adapted to habitual erect posture and bipedal gait; the forelimb is shorter than the hindlimb; the thumb is well developed and fully opposable and the hand is capable not only of a power grip, but at the least, a simple and usually well-developed precision grip; the cranial capacity is . . . (on the average) large relative to body-size and ranges from 600 cc in earlier forms to more than 1600 cc . . . the supra-orbital region of the frontal bone is very variable, ranging from a massive and very salient supra-orbital torus to a complete lack of any supra-orbital projection . . . the facial skeleton ranges from moderately prognathous to orthognathous . . . bony chin may be entirely lacking . . . the dental arcade is evenly rounded . . . the first lower pre-molar is bicuspid . . . The molar teeth are . . . in general small . . .'*

This rather formidable list contains only anatomical features – for the simple reason that it was written by and for palaeoanthropologists studying man's fossil ancestors. These scholars are unwilling to accept as defining characteristics anything for which there is no anatomical evidence. The list can be summarized as follows: man is good with his hands, brainy, and walks upright on two feet; and it is these characteristics that set him apart from all other creatures. Comments on such anatomical features of course ignore the behavioural and cultural characteristics unique to man, in particular speech. In fact many aspects of man's behaviour are the direct consequence of the immense changes to his nervous system that have taken place during the past three million years.

On two feet

During the past 10 million years or more evolution by natural selection has acted upon the human pelvis, and the vertebral column to

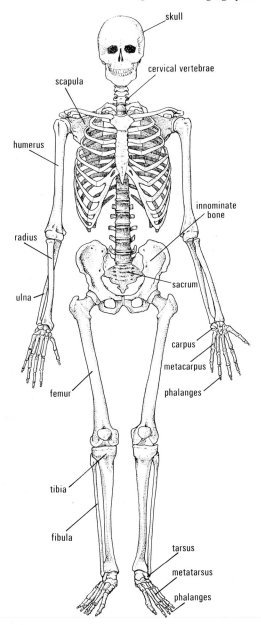

The front view of an articulated human skeleton. Man has adapted to an upright posture, leaving his hands free from locomotion to develop technical skills; he still has the ability to rotate his shoulder blades and forearms like the brachiating or arm-swinging apes.

- skull
- cervical vertebrae
- scapula
- humerus
- innominate bone
- radius
- ulna
- sacrum
- carpus
- metacarpus
- phalanges
- femur
- tibia
- fibula
- tarsus
- metatarsus
- phalanges

which it is attached, to produce a structure that is anatomically unique. From the moment that we climb out of bed in the morning until the procedure is reversed at night we are bearing witness to one of the greatest evolutionary miracles that ever happened: the metamorphosis of a quadrupedal (four-footed) ancestor into a habitually bipedal or upright-moving descendant.

What are the implications of this great change? On the credit side we have become animals with forelimbs (arms) that are relieved of almost all responsibility in helping us move about and which have turned to technology; and with a large brain well equipped to issue complicated instructions to and receive detailed information from these liberated hands. So man has become a master craftsman, but is still humbled by the fact that he is not physically perfect: on the debit side of the miracle are flat feet, piles, and low back pains.

Our ancestors who gradually became more proficient at moving on their hind feet were already pre-adapted to this new method of locomotion. For some of the adaptations that primates acquired as long as 60 million years ago to help them in their tree-living existence proved useful to ground-livers in moving upright. Among these characteristics were the retention by all primates of a relatively unspecialized skeleton (that is, like that of primitive mammals), with mobile ball-and-socket joints at the shoulder and pelvic girdles, unfused forearm and leg bones, and hands and feet with five fingers and toes. The tree-living primates had evolved towards binocular and stereoscopic vision. Above all, there was a predisposition to sit and even move with the trunk of the body held erect from time to time – a tendency that can be seen in primates as different as lemurs and gorillas.

An erect trunk is an integral part of human two-footed walking, as can be seen by comparing man's movements with those of his non-human cousins. Although the latter can hold their trunks erect, because they are quadrupeds or tree-living forms they are not able to

A gymnast hangs from his arms. Like the brachiating primates, he has well-developed muscles for rotating both his shoulder blade and his forearm.

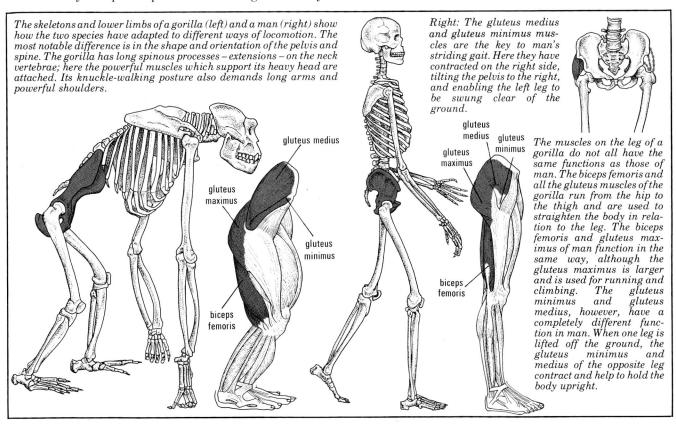

The skeletons and lower limbs of a gorilla (left) and a man (right) show how the two species have adapted to different ways of locomotion. The most notable difference is in the shape and orientation of the pelvis and spine. The gorilla has long spinous processes – extensions – on the neck vertebrae; here the powerful muscles which support its heavy head are attached. Its knuckle-walking posture also demands long arms and powerful shoulders.

gluteus medius

gluteus maximus

gluteus minimus

biceps femoris

gluteus medius

gluteus minimus

gluteus maximus

biceps femoris

Right: The gluteus medius and gluteus minimus muscles are the key to man's striding gait. Here they have contracted on the right side, tilting the pelvis to the right, and enabling the left leg to be swung clear of the ground.

The muscles on the leg of a gorilla do not all have the same functions as those of man. The biceps femoris and all the gluteus muscles of the gorilla run from the hip to the thigh and are used to straighten the body in relation to the leg. The biceps femoris and gluteus maximus of man function in the same way, although the gluteus maximus is larger and is used for running and climbing. The gluteus minimus and gluteus medius, however, have a completely different function in man. When one leg is lifted off the ground, the gluteus minimus and medius of the opposite leg contract and help to hold the body upright.

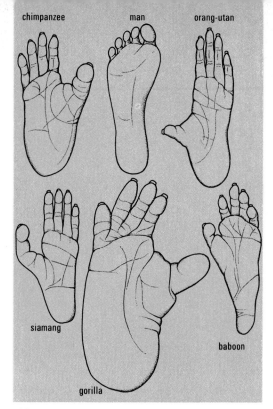

Above: Primate feet. The four apes can all grasp as easily with their toes as with their fingers, because the big toe is opposable. Ground-livers like man and the baboon do not need this ability; the baboon's big toe is only slightly divergent and man's is uniquely non-opposable.

Below: Primate hands compared. The treeshrew has clawed fingers and is ill adapted for grasping. The tarsier can cling even to quite smooth surfaces with the expanded sensitive pads at the tips of its fingers. The chimpanzee has short fingers but an elongated palm making the thumb appear short in comparison. This is even more pronounced in the orang-utan which also has elongated fingers; the gorilla has a long but rather broad palm with short stubby fingers. The gibbon has a generally elongated hand, and the baboon a shortened, broad one. In overall proportion man's hand is a cross between baboon and chimpanzee, relatively broad but with a powerful thumb.

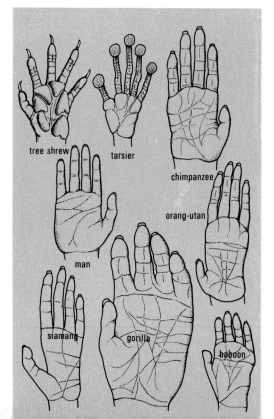

maintain an erect position as general policy. The human spine has all the basic primate characteristics but in an exaggerated form, and has one or two more not shared with any other vertebrate. For example, secondary spinal curves develop in the neck and waist regions when a growing baby begins to sit and stand erect. The fact that the human spine shares many features with the brachiating (arm-swinging) primates indicates that man's ancestors went some way along the road towards adopting this way of moving before coming to the ground and experimenting with walking upright.

The pelvic girdle

Man may be closely related to the great apes, but in the anatomy of the pelvis they are far apart – for the human pelvis has been modified for walking upright whereas that of the anthropoid apes is still that of a four-footed mammal, although they can and occasionally do walk on two feet. But when apes do move on two feet they do so in quite a different way from man; the ape's gait is a crouching, rolling, stiff-kneed shuffle, while a man walks with trunk erect, transferring weight smoothly from one leg to the other and flexing and extending his knees at the appropriate stage of the stride which also involves a heel-toe propulsive movement. In order to cope with man's upright gait the human pelvis has been drastically remodelled during the past 10 million years or so. The whole structure has been markedly shortened from top to bottom, and the upper part (ilium) expanded and widened from front to back. The ilium has been expanded to bring the important gluteus medius and minimus muscles directly above the point where they join the upper end of the femur. These muscles tilt the hip during the stride. In man the gluteus maximus muscle is attached to a greater area of the ilium than in apes; this allows him to remain upright when walking uphill or climbing stairs. Man's iliac crest – a ridge of bone to which are attached the muscles that pass up the back and keep it straight – is also enlarged. His upright position means that the pelvis must bear more weight; to allow for this the area where the base of the spine rests against the ilium is enlarged.

There are also changes in the leg and foot bones. In the leg they are concerned with a shift in the positioning of body weight over the feet and the mechanism of extending and locking the knee. Far more drastic are changes to the foot, which – like the pelvis – is very different from that of any other primate. It retains its primitive mammalian anatomy, with five toes (phalanges), five arch bones (metatarsals), and seven ankle bones (tarsals); but since it has become the weight-bearing platform on which the body rests, foot mobility has had to be sacrificed to stability and strength. So the arch and toe bones have become shorter, and the big toe has come into line with the other toes and acquired a new and immensely important function. For while in apes and many monkeys it forms one half of a powerful grasping organ, being opposable to the other toes, in man it cannot be opposed but forms the launching pad for the body during the final push-off stage in each stride.

Shouldering ahead

It has been suggested that the human shoulder probably evolved from that of a completely tree-living ape, presumably related to the chimpanzee and the gorilla but functionally more similar to that of the tree-living orang-utan. But it would be misleading to suggest that man is directly descended from the extremely long-armed, short-thumbed brachiating apes. It is generally thought now that the ancestors of men and apes diverged before those of the apes acquired these features. Recent research has shown that the human shoulder and arm possess a number of features associated with a more ground-living and four-footed way of life as well as quite a lot that are shared with the upper limb of an arm-swinging traveller. For instance, in four-footed monkeys and man the hindlimb is longer than the forelimb, while in all apes the opposite is true; but on the other hand the muscles that rotate the shoulder blade are well developed in man and the brachiating apes even though man today does not do much arm-swinging. And the muscles that move and rotate the forearm – essential to a brachiator – are as well developed in man as in the arm-swinging apes.

Handy man

At first sight the hand of man differs little from those of the apes. Like almost all primates he has five digits and the sharp claws of primitive mammals have been replaced by flat nails which give increased sensitivity to the fingertips. But there are a number of differences, most of them rather subtle but adding together to make man's hand a uniquely flexible and sensitive instrument.

All the Old World monkeys, apes, and man have fully opposable thumbs – they can be brought into pulp-to-pulp contact with any of the fingers. This pulp-of-thumb to pulp-of-finger grip is known as a 'precision grip'. But man is capable of far greater precision than any other primate. Natural selection has worked on his bones, joints, muscles, and nerves to make his hand the most efficient of its kind in the animal kingdom.

When any of the anthropoid apes plucks an insect from a companion's coat and transfers it to his own mouth he will probably hold his prize between thumb and forefinger, as we would. But his grip will be between the pulp of the thumb and the *side* of the index finger, and the index finger will be more bent than in the human precision grip. For apes' thumbs are considerably shorter in proportion to their fingers than is that of man. Apes have short thumbs because they use arm-swinging as a way of moving about in trees, and a long thumb can be an impediment to a brachiator. The next longest primate thumb to man's is that of the baboon – a ground-liver like man that spends a lot of its time plucking grass or 'manual grazing'.

Although all the anthropoid apes can – and do – swing through the trees with their arms, the gorilla and chimpanzee in particular spend a lot of time on the ground. Out of the trees they 'knuckle-walk' on the backs of their middle phalanges, and to support them their wrist and hand joints are strengthened by ridges of bone. There is no trace of such ridges on man's bones, which suggests that our ancestors probably never were dedicated knuckle-walkers. The tip of man's thumb is markedly broader than that of the apes, and this broadening had started to take place among the inhabitants of Olduvai Gorge nearly two million years ago (see page 57).

So the human hand is functionally superior to that of any other primate – a superiority which stems from the fact that it has never become over-specialized like those of the anthropoid apes, that it has better developed muscles, and above all that it is controlled by an enormously complex nervous system.

The third feature which sets man apart from any other living creature is his large and complex brain. In general form, again, it is not very different from that of the apes, except for its size, and in fact there are various functional similarities between the two. For exam-

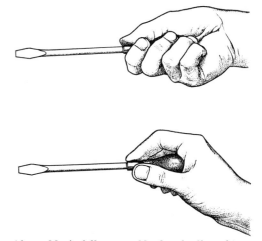

Above: Man's fully opposable thumb allows him to grip effectively whether delicacy of touch or power is required. In the power grip (top) the thumb provides stability and counterpressure to an object held between fingers and palm; in the precision grip the pulp of the thumb is fully opposed to that of any finger, and forms one half of a very sensitive pair of pincers.

Apes have a less sensitive precision grip than man, as their thumb is shorter in relation to their fingers. Above: A chimpanzee prepares to grip a human finger between the pulp of the thumb and the side of the index finger. However, chimpanzees have enough manual dexterity and intelligence to perform complicated tasks like opening locks (below).

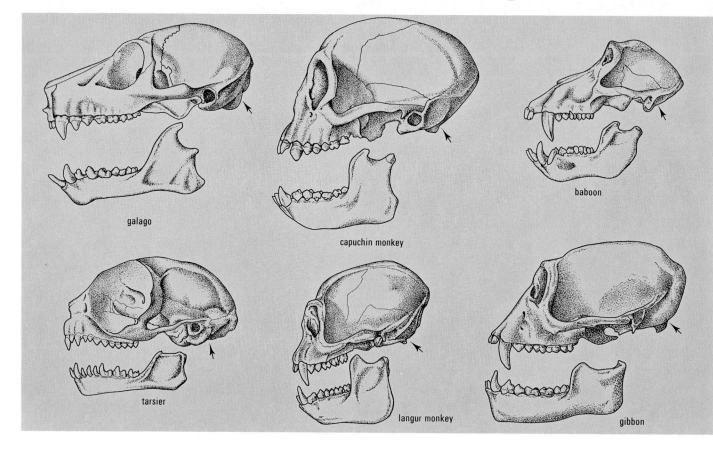

galago

capuchin monkey

baboon

tarsier

langur monkey

gibbon

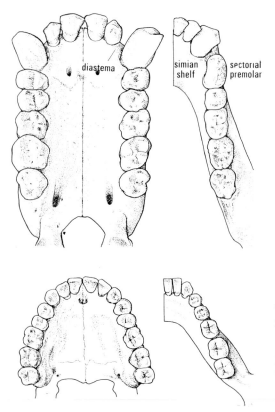

Below: The upper jaw and half the lower jaw of a gorilla (top) and a man. The gorilla's teeth are arranged in a U-shaped pattern; it has huge upper canines and an obvious gap or diastema *between the incisors and canines in the upper jaw.*

diastema

simian shelf

sectorial premolar

ple, the sense of smell is reduced while that of vision is enhanced. Where the human brain differs from that of all other primates is in its handling and storage of incoming sensory information. Little is yet known about the way in which man puts basic urges to good use, but it is thought that the prefrontal areas, which are particularly large in his brain, play an important part.

Man's ability to make use of sensory information is due to the vast increase in the number of connections between one part of his brain and another. Man's association cortex – the part of the brain that receives connections from certain other sensory receiving areas – is much larger than that of the monkeys or apes. The outer cortex of man's brain has a very considerable surface area, much larger than that of other primates, but by a complex of folds – the convolutions – this is fitted compactly into his globular brain box. It is thought that these changes took place after man became upright and began to value his free hands, a combination which was the key to an exciting life. Natural selection favoured those who made the most of it; a large brain was the result.

As the human brain evolved into a larger form, other changes took place in the skull. These are particularly interesting in the study of early man, for so often the fossil remains are of the skull and teeth.

The diagrams above show the many differences between man's skull and those of non-human primates. In particular, the human cranium is globular. An increase in brain size, decrease in jaw and face size, and the adoption of an upright posture have all contributed to this. As the cranium has increased in step with the size of the brain, and as the foramen magnum (the opening where the spine joins the **head) has a more forward position on the skull, the need for large jaw and neck muscles and for bony *sagittal* and *nuchal* crests (ridges on the top and back of the skull) to attach them to, has lessened.**

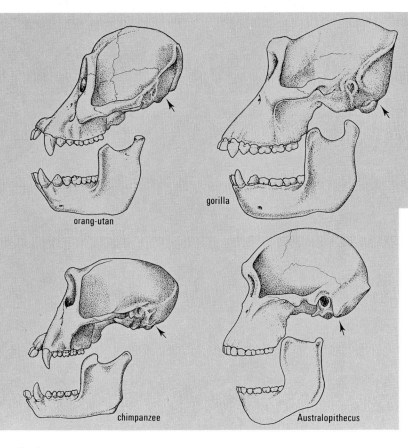

orang-utan

gorilla

chimpanzee

Australopithecus

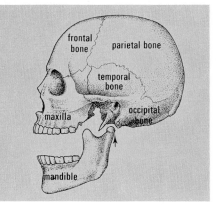

frontal bone

parietal bone

temporal bone

occipital bone

maxilla

mandible

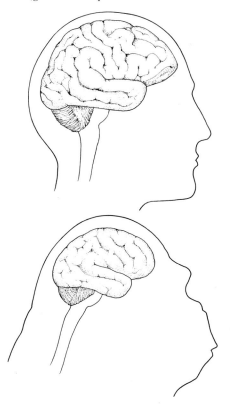

The human face is flatter in profile than that of other primates. His sense of smell has diminished, so he has lost the need for an olfactory snout. Man's ancestors had large jaws with which to tackle their predominantly herbivorous diet; when he took instead to a more omnivorous diet, and to using fire to cook his meat, his teeth diminished in size and his jaws, too, became smaller. Where the facial structure of apes shows heavy buttressing of bone, that of man is less strengthened, producing a more delicately built upper face, and a characteristic chin. His less powerful jaws mean he no longer requires large brows to act as shock absorbers against the pounding formerly meted out to the delicate face bones during strong chewing.

In terms of numbers of teeth present, and the types seen, man is like the anthropoid apes; both have a total of 32 teeth (four incisors, two canines, four premolars, and six molars in each jaw) and the pattern on the raised areas or cusps on the molars is the same. But at first sight ape and human dentitions appear very different. Apes' teeth are arranged in a nearly straight-sided U pattern; they have large forward-projecting upper incisors, tusk-like upper canines that form a pair of shears with the forward premolars of the lower jaw, and lower molars that increase in size from front to rear. Man's teeth form a parabolic curve; his incisors and canines are much smaller and his molars are more regular in size. Although apes and man share a common cusp pattern on their molars, their premolars are significantly different. Man has two very similar bicuspid (two-cusped) premolars on each side of his jaw, both in his upper and his lower teeth. But in the great apes, whether fossil or modern forms, there is a specialized lower first premolar – designed for biting (occluding) against the large upper canine tooth. These important differences show that man and the apes have trodden separate evolutionary paths for quite some time.

Man's Behaviour

During the past few million years, man has changed physically in a number of important respects, including the trebling of his brain size. This selective evolution of a larger and larger brain was clearly related to the demands increasingly made on it. Unfortunately, there is no way of getting from the fossil fragments of the brain box more than a rough impression of the way in which the brain was changing in shape and complexity. Even the relative changes in the regions and convolutions of the brain which are evident cannot be linked with any confidence to changes in behaviour. So the jigsaw of how we have evolved mentally and socially has to be fitted together from a wide variety of different sorts of evidence.

First of all, we can look at our nearest relatives, the other higher primates, and from a close study of their *ethology* – that is, behaviour – we may hope to get some idea of the extent of similar factors in the early hominids. This rather assumes that if brain size is similar in the great apes and

Observations of primate behaviour under controlled conditions can give us an insight into human behaviour. This infant monkey has been separated from its mother and placed in a cage with two dummies, one covered in cloth and the other providing milk. The monkey spent most of its time clinging to the cloth monkey, particularly running to it when frightened; warmth and softness evidently provided more reassurance and comfort.

the Australopithecines, there may have been various similarities in behaviour. Observations of primates have also been important in providing well-controlled laboratory studies essential to a full understanding of our own personalities, but which could not have been carried out on humans. Take for instance the differences in personality resulting from very different degrees of mothering – not the sort of thing which could be undertaken directly on humans, and yet of great importance in understanding later childhood behaviour. A number of infant macaques were reared with different degrees of affection from the mother, and it was demonstrated that this noticeably affected their behaviour later on. In a world in which human antisocial behaviour and violence are all too common, such studies could notably be of immense value in understanding such factors of society, but could also assist in working out possible social action to correct unwanted behaviour. But the trouble with the field of primate study is that it is in many cases all too easy to interpret far more from the evidence than it warrants.

16

Recently the British biologist, Lord Zuckerman, made the following cautionary comment, and it would seem as well, before considering further possible trends in human behaviour through time, to quote him at some length. He writes:

The subject of ethology [behaviour] *has suffered some rude knocks in recent years at the hands of more popular writers who have been successful in conveying not a scientific but a sensational and facile picture of human behaviour. They have encouraged the general reading public to believe that man is some kind of naked ape, obsessed with sex, or a monster unrepentantly acquisitive and aggressive, or a vertebrate whose complex behaviour can be understood if we study the robin, the goose or the baboon. Their message is seductive and simple, and it is not at all surprising that the books they write are best sellers. . . . What is not understood is that much of what they write represents a corruption of scientific knowledge and scientific method.*

What is 'behaviour'?

Animal behaviour, whatever the species, is concerned with what an animal does in its interaction with the environment and other individuals of its own kind. It is concerned with the animal's movement in the environment, its survival in various threatening circumstances, its need to find sufficient food, and the way in which it is usually drawn towards members of its own species for the purpose of reproduction. Unlearned behaviour is called *innate*. A well-known example of this is the way in which the worker bee can indicate to others in the hive a new source of food by a descriptive dance – which codes the nature of the food source, its direction, and the distance from the hive.

In contrast to this – and very important in the higher vertebrates including man – is learned behaviour. Information is stored in a central nervous system, and may be called on to influence further behaviour later. Surprises and problems are always in store for those gathering data on this kind of behaviour. Why do some birds, for instance, produce normal songs even though reared away from others of their species, while other varieties need good contact with their own kind if they are not to produce noticeably abnormal songs? In this case one can conclude that the degree of genetic control versus learned behaviour is variable from group to group.

We learn, of course, in several different ways. The primates are particularly good at gathering useful experience by trial and error methods – without harming themselves in the process. This is an important means of learning in an uncertain or changing environment. Learning by imitation is quicker and safer, and the primates have been good at learning this method. At times even rather novel innovations have been adopted by a primate group, as for instance the washing of items of food by Japanese macaques.

As far as we can tell, learning by *intentional* instruction has always been restricted to man and his ancestors. This is the first part of the cycle leading to cultural development, which in turn stimulates further instruction and behavioural change at a group level. The vastly increased load of things to be learned by human groups has called for the greatest possible extension of the developmental period during which they can be learned. So man's relatively long childhood compared with other animals is an adaptive mechanism essential to the development of advanced hominid cultures.

But with the development of increas-

The human infant remains dependent on its parents far longer than the young of other species. This gives it time to learn steadily the vast amount of knowledge necessary for success in even quite simple human communities.

17

ingly complex societies a dark cloud has appeared, and is only just being appreciated for what it is. For in the development of high-density farming peoples with different levels of nutrition – often grading into states of malnutrition – there is danger of environmental damage to brain development and general level of intelligence. The fact is that brain structure and chemistry are basic to a proper mental performance, and can be affected critically by certain factors of the environment. It has been estimated that today, as a result of severe protein starvation in some communities, as many as 350 million children are at risk and may have retarded personalities as a result. And similar external influences may have slowed down cultural achievements in the past.

The infant in search of a world

Research on the behaviour of infants has grown considerably in the last few years, and some of the findings are important to a proper perspective of human behaviour more generally. As a result of the investigations, many of which use new laboratory techniques, quite a new conception of infancy is emerging – in which a baby is seen more and more as having highly developed perception and sensitivity even at birth. A newborn child evidently has far greater powers of evaluating sensory

information than earlier psychologists realized. At a visual level, there is rapid development, including co-ordinated movement of the eyes and head and sensitivity to colours, and a young baby not only appreciates differences in visual patterns, but clearly prefers certain ones to others. It is likely that some of these early responses may be linked with the sociability of the individual later on. As regards hearing, it has been found that a newborn baby can react differently to pitch, intensity, and duration of sounds, including variation in human speech. All this may well be related to the time at which the child will learn to talk, and have more importance in this than differences in the social environment. In other words, in terms of sound appreciation, we seem to have a considerable built-in capacity for developing language even if the social environment is not ideal. Such detailed work is still in a very elementary stage but it certainly will have wide-ranging value, and may even help in understanding the development of people's anti-social behaviour and aggressive tendencies.

Food and behaviour

During man's evolution he has changed from a wide-ranging hunter and food collector to a highly specialized farmer. At the same time as such food changes great

When threatened, the slow-moving musk oxen form a tight circle, facing outwards, with the young in the middle. This instinctive behaviour was a good protection against wolves but made the oxen such easy prey to men with guns that they nearly became extinct.

changes in behaviour also occurred. Recent studies on such primates as baboons and chimpanzees show that they are capable of killing and eating small mammals. In the case of chimpanzees the male 'hunter' may share the food widely with others. The early meat-eating hominids must have been similar in this respect, perhaps also scavenging meat when available as well. But it is possible that the patterns of hunting and food-sharing of early Pleistocene man tended as never before to show how valuable co-operation could be. More efficient co-operation probably also meant that more meat or larger game animals could be hunted down. The anthropologist Bernard Campbell has listed the following points as being the principal results of changes in the search for food of the earliest hominids:

1. Continued success in hunting required good male co-operation.
2. Nursing and pregnant women would not have had stamina for long hunts, so that home bases of some kind must have been established.
3. Long-distance hunting and meat carrying demanded the evolution of efficient bipedalism – and also perhaps co-operation with carrying poles.
4. Improved hunting required ingenuity in hunting technology and skill.
5. Skinning and jointing problems eventually resulted in the production of better stone tools.
6. Planning the hunt and sharing the food afterwards probably encouraged the development of speech.
7. Accumulating knowledge on animal geography became important, so there was a premium on a good memory and perceptiveness.
8. Hunting probably emphasized the need for a division of labour, with the females attending to all the duties which could be undertaken near the home camp.

Other points could easily be added.

Language becomes essential

Language has evolved as a means of associating some object or concept with a thinkable symbol – most commonly a combination of sounds in a set order. (Initially the sounds are arbitrary, and different peoples have selected different combinations to represent the same thing.) When these words are combined to give sentences – strings of thought – they can transmit considerable information from one person to another. The most important effect of language on the cultural evolution of man is that it both increases the learning capacity of people, and makes learning itself quicker.

The elaboration of language in man is really an extension of the sound-signalling or 'call system' seen generally in birds and mammals. These signals may act as a warning when danger is near, establish mating or territorial rights, be concerned with mother-child relationships, announce the appearance of food, or emphasize the workings of the social structure. Certainly it is doubtful whether monkeys and apes would have become as intensely social as they are without the great range of sounds they produce. But there is a considerable difference between a call system and a developing language; calls are 'closed' systems, unyielding in expression, while language is versatile and 'open', allowing new combinations of sounds when needed. A call is simply linked to a direct stimulus, while language can be concerned with things past, present, or future. Calls are mainly inborn, but language as a complex whole must be learned, and must be memorized and spoken from generation to generation.

The evolution of language has apparently dictated genetic and structural evolutionary changes as well. Speech is certainly concerned with certain areas of the cerebral cortex of the brain which are poorly if at all developed in other primates. When this expansion took place is debatable, but certainly by Middle Pleistocene times the brain size of hominids had increased enough to suggest that the speech centres were part of the expansion.

Family and kinship

The human family unit, where there may only be one senior male, is an unusual state in the primates, although in some other vertebrates there are instances where the father supports and protects the mother and her young. Social scientists think that this basic group in man has been determined by economic and political needs rather than being based on any special biological value.

In most species the young grow quickly and leave the primary unit, but human children are unique in remaining with parents over many years since it takes them such a long time to reach maturity. The male is necessary to provide for them and their mother while they are dependant. One of the few alternatives to this situation is *polyandry*, where instead of one father in a group, the woman is permanently married to more than one man. It has been described in the Todas of southern India but is a rare alternative; it may possibly be linked to a shortage of women caused by the killing of baby girls as a means of population control. There is no reason to think that it was any more common in the past. But *polygamy* or marriage to more than one wife has certainly been common enough in some parts of the world in recent times, and there is literary evidence to suggest that it probably has a long history. In this form of marriage, too, the slightly more complex family structure is justified by economic factors, including a bride price.

The generally exclusive nature of sexual relations between male and female in marriage is likely to be related to the links between father and children in terms of economic liability. A social contract of this sort is something of a guarantee for the mother and children of economic support,

and it also has the important function of welding together different descent or tribal groups.

One can only guess at the situation in the early hominids, whatever comparative information we have on other primates and modern groups. It is at least probable that one-male groups evolved early, and were ideal where hunting and collecting in an area would not support many families. With the development of co-operation, the males from several families would join in collective hunting groups. We can surmise that with the gradual increase in the numbers of hominids during earlier Pleistocene times there would be increasing demands on territory for hunting purposes, and probably this was a source of conflict on many occasions. It could well have encouraged the political strategy of

A hunting party of South American Indians. The study of societies hardly influenced by civilization can help us to draw conclusions about the probable way of life of early men.

exogamy – marriage outside the kin group. In primates generally, endogamy – with mating only within the group – is the common pattern. Exogamy which established ties between groups was therefore an important innovation. Its practice was encouraged by the growth of rules concerning incest – mating between individuals who are very closely related. In turn this can only be achieved by recognizing degrees of kinship – father/daughter, brother/sister, first cousins, and so on.

Kinship must also have been seen in terms of bonds of affection – a feeling which is well developed in man. This would also have meant the recognition of obligations to the older members of the kinship group, and it is interesting to recall the 'old man of La Chapelle aux Saints', a Neanderthal individual with poor dental health and very severe rheumatism, who may well have been cared for by more active members of his family. In other animals the old and sick, who cannot keep up with the group, are left behind to die.

Built into these increasingly complex relationships was the increasing human ability to control the situation by conscious thought. This included sexual behaviour, which became more and more liberated from pure physiological drives. The final outcome of gradual population increase and the elaboration of kinship ties was the development of tribal-size units and eventually whole nations.

Aggression and war

Behaviour between groups of non-human primates is usually unfriendly and antagonistic. Some people believe that man also has this inbuilt aggressiveness, and it is easy enough to think this must be so when one considers that in the First World War of 1914–1918 8 million men were slaughtered on the battlefield, and possibly another 15 million people died as a result of the war. Sir Arthur Keith, the anatomist, saw such activities as 'nature's mechanisms for preserving the individual and the tribe or nation'. But can we, on the evidence of modern behavioural studies, still justify this viewpoint?

First of all, it should be noted that violence may result from aggression, but that aggression is a more basic social phenomenon, used as a means of achieving one's own ends or because the person or group feels threatened – but without real provocation. The aggressive reaction may be linked with territory problems, differences in hierarchy relationships, or to frustrating circumstances. In primate groups generally, such conflict could occur over food resource problems, between young animals in play, or as threats to positions of dominance in the older males.

Some primatologists and sociologists have been particularly interested in the question of male dominance, and the extent to which it has played a part in our evolution – and perhaps still does. They argue that some form of dominance hierarchy in a group has been essential to the maintenance of social cohesion and organization. In such a state, dominant males have a breeding advantage over the others – good for the group as a whole. This demands of the subordinate males some inhibition as regards aggression towards the dominant males. With the evolution of artefacts which could be used as weapons, the situation probably became increasingly complex, and may in turn have spurred on the further evolution of the brain to cope with these further social changes. Loyalties to other males, varying actions of a competitive or aggressive kind, even perhaps developing feelings of guilt when individual action threatened social peace, can perhaps be seen as the outcome of this basic dominance phenomenon associated with human cultural development.

A male baboon confronts an intruder; the next step is the bared teeth of aggression. The ground-dwelling baboons must constantly be on guard against predators, and are therefore among the most suspicious and aggressive of monkeys. They have developed a complex group structure based on a hierarchy of dominance.

How does man's aggressive behaviour compare with that of the primates? One important difference is that when combat has occurred in the hominids it is always likely to have been of a very different form to that of primates generally, who use their large canine teeth in fighting. Biting has not been man's principle method, but because of his bipedalism, his hands have assumed great importance – being particularly dangerous when armed with stone, bone, or wooden tools. Another difference is that while primate groups generally show aggression to one another, studies on so-called primitive communities made over the last century produce as much evidence for the generally peaceful nature of many such peoples as for any particular aggressive behaviour. On available evidence it can be argued that the hominids only enter into really severe forms of conflict as a result of population stress – a subtle and malignant pressure – or perhaps when driven by economic and political factors. Certainly, the terrible destruction wrought by man against man during historic times must be viewed as states of social pathology, not typical of human populations during the many thousand years of man's evolution.

Although much has been written on territoriality in some animals, the extent to which primates – and especially man – think territorially is still a very debatable subject. Since the emergence of *Homo erectus*, the hominids have been increasing in numbers and moving out into new lands. In the Old World, hunting groups might perhaps have been settling down to roughly defined territories for exploitation

by a hundred thousand years ago. Further expansion of particular cultural 'empires' might then have been possible only in stronger competition with already established groups. So population and territory needs for survival may have emphasized the value of establishing firmly claims to territory, and this in turn would have called for the further elaboration of political machinery needed to carry this through. Defence of territory may well have a long history, well back into Pleistocene times.

Ritual and religion

Unlike animal rituals, those of humans rarely relate to pressing biological needs – but they are types of community signals nevertheless. Human rituals are in fact stereotyped behaviour patterns, consciously enacted by the individuals in a group – and with set words and actions. The meaning for such activity is not easy to analyse, but there seems little doubt that the development of hominid ritual is bound up with the perpetuation of certain kinds of knowledge of long-term survival value to the community. In other words, it is another kind of language which can be interpreted only within the particular social context. It has been suggested that ritual was to some extent earlier in hominid cultural evolution than language – a first means as it were to take note of human achievements. With the event of language and later of writing, myths were developed as an attempt to record some of the ritual content of a community.

Where rituals bind the community together as a whole, they might be regarded as religious. The more communal effort and skill is needed in performing these religious functions, the more the individuals in the group are likely to feel a closeness with their fellows. This is by no means the only interpretation which can be given, and Desmond Morris, for instance, sees religious activities as a coming together of large groups of people to perform repeated and prolonged submissive displays to appease a dominant individual. But religion has clearly far more to it than that. We have only to think of the Ten Commandments to appreciate that religion also emphasizes worthwhile laws in a society.

Religion, like other major aspects of human society, has adaptive value – it

Far left: A group of 'whirling dervishes', whose dance forms a major part of their religious ritual.

Much ritual has grown up around burial ceremonies, partly as a way of coming to terms with death. Here Chinese dancers and musicians take part in a traditional ceremony by a grave to exorcise demons.

A Thracian marble idol, dating from about 4000 BC and showing the head and torso of a woman. Most of the Thracian idols depict a mother goddess; myths of such earth mothers, often very similar, are widespread in the religions of early societies and are often connected with fertility and harvest.

helps man to come to grips with his fellow men and with the world around him. Whether magic, witchcraft, ancestor worship, or reverence for high gods, it is a part of the struggle for existence, part of the survival plan for each community. Of course myths and rituals may be carried through the generations long after they cease to serve a useful purpose, but this does not alter the fact that they initially developed in a society to fulfil a need. The major religions – Christianity, Buddhism, **Islam, and so on – are not so easy to** analyse as those of so-called primitive societies because they have developed into large and complex philosophies on life which now cut across various societies in different countries. At times, however, remarkable similarities in action and human response can be seen, even when comparing aspects of 'primitive' and 'advanced' religion.

Basic to all religious activity is really man's awareness of the uncertainties of the universe around him – and even today science by no means knows all the answers. When this sensitivity to the unknown evolved in the hominids we do not know. Certainly by half a million years ago our brain size might have permitted the beginnings of doubts and fears beyond those experienced by other mammals, but concrete evidence is not forthcoming until the ritual burials of Upper Pleistocene times.

Primates such as chimpanzees have been observed to show considerable distress and even long-term depression as a result of the death of a close relative, but man alone has contemplated death and reacted to it at a religious level. Burial customs certainly go back at least 40,000 years, some Neanderthalers having been buried with great ceremony. It is a short step between thinking of the recently dead and those remembered from long ago. And ancestor worship has provided further ritual behaviour which clearly had value in binding the group together. There is thus in rituals – even those which at first sight seem exotic and valueless – a strong underlying factor of reason which is geared to the wellbeing and survival of the group as a whole.

Visual art and music

The evidence of upper palaeolithic cave art (see page 86) shows that man has been able to produce fine works of art for many thousands of years. But in this late Pleistocene art we see fully developed artistic abilities, and such evidence really gives us no idea of the true antiquity of artistic interest in man. That the great apes have produced 'pictures' in the artificial environments of zoos and laboratories does not help too much in finding a beginning for real artistic aspiration in earlier hominids. Perhaps body decoration related to ritual was the earliest form of art – beyond the level of shaping tools into particular forms. In any case, the size of the hominid brain by Middle Pleistocene times would certainly not exclude the appropriate level of perceptiveness for the development of experimental art.

In the case of music, instruments are not found until late in our prehistory, and yet this may have been an important aspect of much earlier cultural development. Charles Darwin put forward the view that the music of earlier man was a development from the cries and drumming actions of the kind known in apes. But Leonard Williams, a musician who has also made a study of primate behaviour, disagrees. His alternative, which seems more reasonable, is that music is basically rhythmic rather than melodic. Again there may be a link into developing ritual, and like visual art music perhaps began as a reinforcement to other social needs.

Man and his Environment

The environments enveloping man on the Earth's surface form an intricate spider's web of factors which he cannot ignore in his day-to-day business of life. He must cope with the weather, eat foods derived from the biological world around him, keep warm, use metal in ploughing or driving to the office, and so on. In just the same sorts of basic ways men from the earliest times have similarly been enmeshed in, and exploited, the environment around them.

What does this web surrounding human groups consist of? First, of course, there is the place itself – one or other part of the habitable world, be it fertile valley, hillside, semi-desert, or high in a mountainous region. To some extent linked with these factors of latitude and altitude are those of temperature, humidity, and even windiness. Such things can either favour human settlement or work against it. Where some factor of an environment can be viewed as detrimental to man, and yet despite this he remains there, then there is sure to be some adaptive reason for it. Man can change physically or by means of his culture to suit his environment. In some parts of the world, as a result of bright

An oasis in the Ethiopian desert. Such sources of water in generally arid areas have been essential to the survival of man and his domestic animals.

sunlight and especially of ultra-violet rays on his skin, there has been the gradual evolution of darker and darker skin colour. For extra pigment protects the deeper regions of the body from such harmful radiation. A different kind of adaptation is the creation of the well-designed slit goggles of some Eskimo communities – an excellent protection against sun glare reflecting off expanses of snow.

Among basic factors of the environment, water is an essential for human existence, and thus even in the case of the primitive nomadic groups there was a continual need to return to rivers or water holes. When eventually some peoples began to settle, proximity to drinkable water was clearly an important factor. But at times danger lurked in the water – disease to be handed on to the unsuspecting humans. This was particularly prevalent in certain areas, and perhaps has been a major factor in keeping down the numbers of some earlier African peoples since water in Africa may contain micro-organisms causing such diseases as dysentery, cholera, and infective hepatitis, as well as being indirectly linked with the transmission of bilharzia, caused by parasitic worms.

Since earliest tool-using times, the hominids have had some awareness of the natural inorganic resources of the Earth's crust. At first they made use of rock for crude hand tools and then of the more easily available mineral ores; now drilling

25

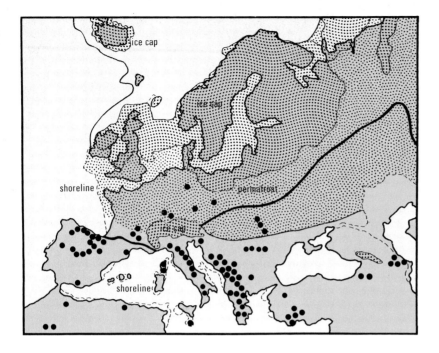

lowering of the sea-level, and the massiveness of these great areas of ice resulted in zones of cold arctic conditions immediately south of the ice. Even Africa, much farther to the south than the limit of the ice sheets, saw at least some climatic change, and in some more humid periods Mediterranean conditions prevailed in areas which are now desert.

With the final retreat of the ice came the so-called postglacial period, although it could be argued that perhaps the present mild world climate is simply because we are in an in-between or interglacial phase – with more ice to come in some hundreds or thousands of years from now.

Rocks and soils

The Earth's crust forms a relatively thin layer on its surface – the rock thickness

great depths into hundreds of feet of strata permits man to search for coal, diamonds, oil, and even radioactive rocks of possible importance in producing atomic energy. Human groups, especially when farming had evolved, were also to be considerably influenced by the thin outer layer of the Earth – the soil. It is the soil which controls the nature of the vegetation, and is thus ultimately linked with the animal life which the plants directly or indirectly support, and which in turn support man. All these factors have been influential in man's past development and progress.

Climate

On present fossil evidence, it seems likely that the earliest home of the hominids was in more tropical areas, for until the discovery of firemaking some half a million years ago, their higher temperatures and humidity and periodic heavy rains were more suited to human survival than colder, windy, and damp climates.

The Earth's climate has been by no means constant during Pleistocene and more recent times. A series of four major ice ages or glaciations took place and during these times vast tracts of ice covered areas of Europe, Asia, and the Americas which are now quite free of it. Examples of the same kind of expanding and contracting ice sheets can be seen today in Greenland and Antarctica. This locking away of so much water in thick ice sheets caused a

Above: The last Ice Age at its greatest extent. The area of land was generally greater than it is today, but the Black Sea was larger. The ice cap covered all of Scandinavia and most of Britain; Northern and much of Central Europe were an area of permafrost – soil that is always frozen hard. Small glaciers covered the Alps and isolated mountains (indicated by black dots). There were forests in Central Europe, but north of the black line on the map no trees grew.

Right: A Solutrean laurel-leaf blade, delicately flaked from flint. It was a characteristic part of the stone tool kit of some upper palaeolithic people of France and Spain. From much earlier times man's pre-human ancestors learned to shape tools from the stones he found lying around; later man learned how to mine flints.

can be compared to the thickness of a postage stamp on a football – yet the rocks and minerals still provide vast resources for man to exploit. The three major classes of rocks are igneous, sedimentary, and metamorphic. All have been used and mined. The igneous rocks have crystallized from a molten state and are especially strong. As such, they have been useful for a variety of tools, from axes to quern stones. The Egyptian pyramid builders at times selected this hard stone when especially strong supports were needed. It also polishes well, so that statues and other art were made from such rock.

Sedimentary rocks are, as the name implies, derived from sediments – usually layers of decomposed igneous rock. Also included here are various deposits derived from chemical precipitates. Salt is one of these, and has been of growing economic importance to man since at least neolithic times. Flint, a concentration of silica within chalk rock, was especially important to Pleistocene man for tool-making. Some of these sedimentary rocks are fine grained and can be worked into vases and so forth – for example Egyptian alabaster. Early builders were quick to see the value of some of these materials. Gypsum was used to make house floors in ancient Jericho, and bitumen lined certain walls in the early city of Mohenjo-Daro built by the Indus valley peoples.

Of least importance to man, perhaps,

are the metamorphic rocks, which result from the other two types being influenced by great pressure or heat; most used in early historic times was marble, which is a recrystallized limestone.

The Earth's crust also contains metals which Man has learned to exploit. The progress of early metallurgy was generally quite slow, but was nevertheless critical to the development of urban civilizations. The sequence of metal exploitation was by no means the same in all parts of the world, although in Europe it was established as a Copper Age leading to a Bronze Age, then to an Iron Age. In Africa, however, the use of iron preceded that of bronze. The fundamental stages of metal use were first the working of native metals – copper, gold, silver, and meteoric iron – by hammering and cutting; secondly came the exploitation of metal ores and the production of more refined metal or alloys – lead, silver, copper, tin, and bronze.

More and more studies are being made of man's exploitation of the rocks and minerals around him, and sometimes these lead to quite detailed detective work. For instance, it has been possible to make detailed studies of stone axes found in different parts of Britain and actually identify some localities from which they came. This not only tells us something about stone axe factories, but also about the development of prehistoric trading.

The quality of soil must slowly have come under the consideration of man for other reasons than its mineral content. Soil may be defined as a thin layer of material over much of the world's land surface, which is initially derived from the decomposition of underlying rock but is mixed with decayed plant and animal material and contains microscopic life as well as larger organisms such as earthworms. Its importance to man has been in the extent

27

to which it could support useful vegetation, and this became particularly important with the development of agriculture. Cultivation, especially with the limitations of early ploughs and digging equipment, was clearly best in a moderate climate where the soil was fertile but not too heavy. It is therefore no wonder that the lower valleys of certain rivers in the Near East such as the Nile, Tigris, and Euphrates, with deep rich soil, were among the places where early farming settlements grew up, and where the extremely fertile soil allowed crop surpluses to support the first cities (see page 109). Thousands of years later, it was the distribution of coal and mineral ores that led to the concentration of industrial towns.

Plants

Vegetation varies considerably over the world's surface, and has been vital to the survival of man – not only directly by supplying fruit, nuts, seeds, and roots, but in maintaining the animal communities he hunted. It shows great variation, from lush tropical vegetation to semi-desert scrub. Within one region plants can also show clear preferences for slightly different conditions: for instance willow is water-loving, beech and ash prefer calcareous soils, and rhododendron lime-free acid conditions.

As climates have changed through the period of human history, so has the nature

Above: The first cities grew up in the fertile river valleys like that of the Nile, where the rich soil allowed farmers to produce crop surpluses to support craftsmen and administrators. This part of a wall-painting from the tomb of Meruruka (early third millenium BC) shows ploughing (top); loading sheaves on donkeys (second row); reaping to pipe music (third row); and treading the grain (bottom).
Below: Careless farming has led to disastrous soil erosion when vegetative cover has been removed. Modern farming tries to prevent this by methods such as contour farming, which can stop soil washing straight down hillsides.

of the vegetational cover become modified. A marked change can be observed in the occurrence of some species of plants in a particular locality through time, and there is now a mass of information on such fluctuations from the last part of the Pleistocene through into postglacial times. Much of the plant evidence comes from detailed studies of pollen grains, which are well preserved in some deposits and show great shape variation depending on species. We can begin to link up, then, a history of floral change with the history of man, and in some cases the two are closely involved. The greatest plant changes have come about as a result of man's continued development of agriculture, which has demanded much forest clearance with ever-increasing areas of grassland. Timber has also been in demand as fuel, for house building, and for defences and fencing. It is therefore small wonder that in Britain, for example, there are no large old forested areas remaining intact. Man's destruction of vegetative cover and careless farming practices have led to the devastation of huge areas, such as the 'dust bowl' of the south-western USA.

But the early farmers were to have further impact on the plant world, for they saw the value in selecting plants for cultivation, and in the end greatly changed the shape of some Old World cereals, maize, tomato, and a variety of other plants.

Animals

Most of the food animals of importance to earlier human groups were herbivores

(plant-eaters) – a further link in the environmental web into which man fits. But this is not to say that palaeolithic hunters did not exploit every possible animal. The Neanderthal hunters of Teshik Tash in Russia killed brown bear, hyaena, and leopard, while prehistoric people of Niah in Borneo included in their meals no less than 11 carnivores. It seems likely that this continuous quest on the part of earlier man for animal foods has had serious repercussions on the survival of a large number of mammal and other vertebrate species, and by the end of the Pleistocene period human over-killing had caused many to become extinct, such as the mammoth and giant beaver. This may have been an important contributing factor in the domestication of certain animals – as an insurance policy against further decline in wild varieties. This was only one factor, however, for some domestic mammals we keep today – especially the dog and pig – may really have adopted man initially as a source of food scraps. It is also possible that cattle were first controlled not for their food value, but for certain ritual reasons – and this may also have been the case with the breeding of cats in early Egyptian societies as well.

Studies on certain species of animal have also provided information on other aspects of the environment, both climate and habitat. A good contrast is seen for instance between certain mammals in Europe during the Riss glaciation and an interglacial period. In the former there were cold snow steppe conditions with woolly mammoth, reindeer, and woolly rhinoceros; in the latter warm forest, and parkland animals such as the straight-tusked elephant, bison, deer, and a larger species of monkey. In the past few decades there has been a growing interest in invertebrate animals from the point of view of further environmental reconstruction – especially molluscs and insects. As yet the evidence is very restricted, but there is no doubt that these areas of scientific investigation in prehistoric studies will provide much more information in the future. By late Pleistocene times at least, skills and equipment for catching animals had increased considerably. Mesolithic man was probably far better able to fish inland waters and the coastal seas. His knowledge of wild fowling probably brought him into far greater contact with birds and no doubt this knowledge of their behaviour was valuable when eventually certain birds – barnyard fowl, ducks, and geese – were domesticated. In coastal areas some mesolithic and later communities made good use of other sea food, and especially shellfish. In various parts of the world shell middens mark the sites of numerous meals which local groups made of molluscs. Eventually, in Roman times, even snails and oysters were to be valued sufficiently as food to be controlled – semi-domesticated – as a delicacy for the tables of the upper classes.

Disease and man

It is usual to think of the environment in relation to earlier man from the point of view of its exploitation, or the limits climate may have placed on his settlement. But the last word should perhaps go to another factor, that of disease. For human survival has also meant a constant battle with a wide variety of micro-organisms, as well as larger parasites such as intestinal worms. Although we still know little about this part of our history, we should not ignore its importance as a part of the study of our environment, or the fact that it has probably undergone great change. In particular, the gathering together of people into towns and villages enabled epidemic disease to survive and strike as never before. The diseases, then, which we see in world communities today, have a long history, no shorter in fact than that of the rise of man himself.

Neolithic carvings from the Algerian Sahara show a leopard and an ostrich – evidence of a changing environment since such animals need a much more fertile habitat than that of the region today.

In order to measure the direction and intensity of the Earth's magnetic field at some date in the distant past, it is necessary to collect a sample of burnt clay or stone. This has the information 'frozen in'. The sample is capped or enclosed in plaster and its orientation is accurately marked before it is removed from the ground. The direction of the ancient field may then be discovered from this alignment, and the information put to use for dating future discoveries (see page 32).

Science Investigates Prehistory

As recently as the middle of the last century there were still educated people who believed that the world was created in 4004 BC, and that mankind had been on the Earth from its beginning. This view was based on a calculation performed by Archbishop Ussher on chronological data found in the Bible, and published by him in 1650, at which time it was widely accepted. Ussher was by no means the only person to have made such an estimate, and many of the world's religious doctrines had attempted to place the position of man in time, with some quite remarkable variations.

Such ideas as to the relatively recent origin of the Earth and its inhabitants were not made in the complete absence of any evidence to the contrary. Fossils had been collected and treasured as ornaments by prehistoric man, and certainly it was known by the more learned towards the close of the Middle Ages that these were the remains of ancient living creatures and not, as was popularly thought, objects manufactured in imitation of living forms or the petrified remains of recent life. But the usual explanation for this – and that most acceptable from a theological point of view – was that fossils represented the creatures destroyed by the Great Flood of Genesis.

During the 17th and 18th centuries, ideas of the geology of the Earth began to take shape. In about 1603 an Englishman, George Owen, had observed and made some suggestions as to the nature of strata, and these were developed further by Nicolaus Steno in Denmark a century later. A clergyman called Thomas Burnet published his *Theory of the Earth* in 1681, in which he described the alterations which had taken place in the Earth's surface; and while he still attributed these largely to the influence of the biblical Flood he nevertheless realized that the timescale involved must be considerably greater than that traditionally accepted. At about the same time another clergyman and distinguished naturalist began to examine a number of fossils. This was John Ray, and he established that these examples contained not only shallow water species, but some from deep waters and others which were apparently not members of known and surviving families. This strengthened suspicions that geological timescales were in fact very long, but it was not at that time possible to determine the actual period involved although Lamarck in 1809 made an inspired guess that the true age of the Earth must number thousands of millions of years.

Measuring time

The techniques used to measure geological time today are in many cases the same as those used for archaeological measurement – that of the short portion of the Earth's history which has been seen by man. It is important that the age of the Earth and the succession of geological time be established in order to determine man's place in nature and in relation to other living creatures. For this reason much work has been carried out during the last hundred years, utilizing a variety of scientific techniques for the purpose. From these methods two kinds of date can be obtained: the so-called *absolute* date, which gives in years the time which has elapsed since a particular event, and the *relative* dates, which tell only how one period or event stands in relation to another. Both are of importance to the geologist and the archaeologist, but obviously it is particularly interesting to acquire absolute dates where this can be done.

Among the methods tried in the 19th century were determinations of the salinity of the seas and the rate of addition of salt from river waters, and rates of erosion and sedimentation. The first real attempt based on exact physical laws was that of Lord Kelvin, who started with the current assumption that the Earth was formed from matter torn from the Sun, and had thus originally been at the Sun's temperature, and utilized Newton's law of cooling. This gave the extraordinarily low age of 40 million years, which excited scientific controversy for 50 years, but was qualified by the cautious statement of this eminent physicist that it would only be true 'provided that no new source of heat was discovered'. That prophecy of course came true with the discovery of radioactivity, but this apparent complication in fact provided the basis for a number of the most important scientific dating methods. As the result of the application of certain of these techniques to the study of the Earth's oldest rocks, it appears that they are probably more than 4500 million years old, with the Earth itself up to half as old again. Of this vast time span, it is in fact the last five million or so years we are principally concerned with, as this is

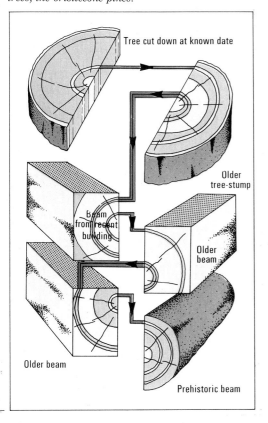

A tree-ring chronology (see page 33) is constructed by matching overlapping rings of successively older wood samples. When this has been done, new samples from archaeological sites can be dated by finding their position in the ring sequence. This method of dating has been of great use in the south-western USA, where a climate ideal for the preservation of wood is combined with the presence of the world's oldest living trees, the bristlecone pines.

Tree cut down at known date

Older tree-stump

Beam from recent building

Older beam

Older beam

Prehistoric beam

Above: A sample of clay showing the sequence of layers or varves formed by the deposition of fine particles at the edges of lakes formed by melting ice. The level of such lakes would rise each year by a slightly different amount, causing a pattern that can be followed and used for dating.

Pollen grains preserved in soil samples can identify the plants that once grew on a site, and so be a clue to whether the climate has changed. The pollen of each species has a distinctive shape when seen under the microscope. The chart below is used to recognize some common types. Bottom: A soil sample from a Bronze Age barrow in Hampshire, England, with pollen grains including grass (3) and lime (7).

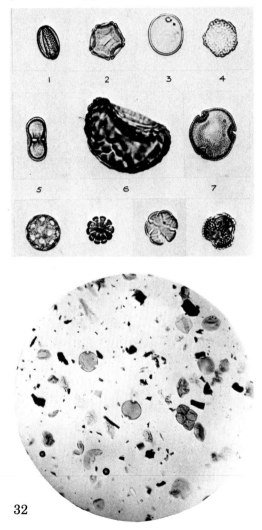

the period for which there is at present clear evidence of man's existence on Earth.

In order to date the various phases of man's development, several different approaches are possible. A site in which remains are found can be dated by means of some geological feature which can be recognized elsewhere, or the remains of animals or plants found there can be identified and interpreted in terms of the fauna and flora present and the climate which this would represent. This may then be compared with sequences of similar material from other sites and provide a position in the relative timescale of such events. Procedures of this type are known as relative dating techniques, and the actual time in years may not even approximately be known. Alternatively, one may utilize one or more of the absolute dating methods to date objects or materials found on the site, and so obtain a date in years.

Comparing the evidence

Almost any feature of the environment which changes with time and which leaves some record in the ground can be used to set up a sequence of relative dates. For example changes in shoreline or river level, caused by sea-level variations resulting from the melting of ice at the close of different glacial periods, leave recognizable traces. Events at different sites which are detected in the same river terrace can therefore be assumed to be of similar date, and from a series of such observations a sequence can be built up. It may be possible to obtain by other means an actual date for a river terrace or shoreline, and if that can be done a date can be assigned to any further material from that context.

As early as 1858 the French palaeontologist Lartet had shown that it was possible to identify different ages over an area by looking at the presence or absence of the remains of different mammals, and this has proved extremely useful for dating the deposits of material found in caves and brought there by man and animals. It is not always particularly accurate, as some species survived in small pockets when they had died out elsewhere, but it provides a useful guide.

In addition to mammals, the insect, mollusc, and other animal remains found on sites have been utilized, but in recent years much very valuable evidence has been obtained from one special type of plant material. The technique of pollen grain analysis consists of the separation of fossil pollen grains from certain kinds of soil of which peat is perhaps the best, and the identification of the species of plant which they represent. This gives a very useful picture of the kind of vegetation on and around the site at the date corresponding to the level from which the soil was taken, and also allows the prevailing climate at the time to be discovered. These pollen grains are, of course, exceedingly small, but they have widely varying shapes which are characteristic of the plants from which they come. Once again, study of the relative numbers of different pollen grains from different levels in a single site allows one to build up a sequence which reflects the changing plant population and climate of the site, and this sequence can then be matched with others from different sites to find to what extent they are contemporary.

'Frozen' north

A physical aspect of the Earth which varies with time is its magnetic field, and this variation over recent years is readily observed by comparing the angles between true north and magnetic north over that time. Fortunately we have some record of the variation in this magnetic field in the more distant past. Many if not most rocks contain traces of iron minerals which act as minute magnets. During the formation of the rock, or the solidification of a lava flow, these magnets have an opportunity to become lined up with the Earth's magnetic field and 'frozen' into position. If in the intervening period between that event and the present day the rock does not become moved, it will retain a 'memory' of the direction and strength of the Earth's magnetic field when it was formed, and the direction of magnetic north at that time can be established by careful measurement. Two rock samples showing the same geomagnetism, as it is called, may therefore have been formed at the same date, although it is, of course, possible that the field pointed in the same direction on more than one occasion.

This technique can be applied to the dating of archaeological sites, for not only can natural events such as volcanic eruptions cause the formation of these natural fossil compasses, but the lighting of fires by early man may fire the mud into a clay which becomes magnetized in much the same way. Portable objects like pottery which have been fired also retain such a magnetic memory, but because they have **almost always been removed from the position of firing** their information is rather more difficult to decipher. When sufficient measurements have been made of the magnetic direction and intensity of material from sites of known date from a particular part of the world, a map can be constructed which shows the movement of the Earth's magnetic north pole, and which can with care be used to date other sites in the same region.

Overlapping rings

Another technique exists for dating ancient sites for which no sophisticated equipment or ideas are necessary. Wood exhibits a series of concentric rings, and these are the result of the annual growth of the tree. Well over a century ago Charles Babbage recognized that these rings varied in width from year to year as a result of the effect of climatic variations on the growth rate. He proposed that this variation might be used to identify the range of years over which a tree had grown, and hence be used for dating the buildings into which it was incorporated. This led Professor A. E. Douglass in the 1920s to search for ancient trees which could be used as markers in the development of a long timescale based on tree rings. The reason for this was twofold, for not only was it possible to compare the rings from different samples of wood to see whether they matched, in which case the trees from which the wood had been obtained were growing for at least a portion of their lives at the same time, but it was also feasible to obtain an actual date in years for any particular ring. This was done by taking a sample from a living tree (normally with a long thin boring device which removed a core of wood) and counting backwards from the present year ring by ring. If the tree was old enough it was possible, at least in theory, to find the period during which any other tree was growing by simply matching the sets of rings with those of the dated sample. This was obviously much easier if the tree used for reference purposes was as ancient as possible, and hence Professor Douglass's quest. It was possible to construct a long timescale without using the most ancient trees, but again one had to start with a living tree. This was matched not with another tree but with something known to be rather older, like a beam from an old building or the remains of a tree stump. With luck some of the earlier rings from the tree would match the later rings on the dead wood

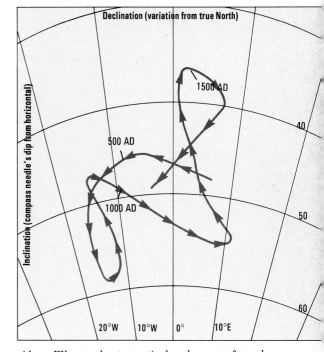

Above: Where rocks at a particular place were formed at known dates (for example by volcanic action) a graph can be plotted showing the variation in the Earth's magnetic field at that point. This graph is based on lava flows from Mount Etna, whose eruptions have been reliably recorded for an unusually long time, and gives readings for two thousand years.

The bristlecone pine, Pinus aristata, *which grows in the White Mountains of California, is thought to be the world's longest-living tree. It has often been used in tree-ring studies, and can provide information about climate as long ago as 4000 years.*

33

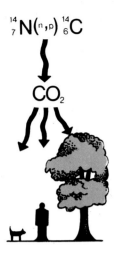

$$^{14}_{7}N(n,p)^{14}_{6}C$$

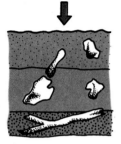

Natural radiocarbon is formed in the upper atmosphere by the cosmic ray bombardment of nitrogen atoms.

This is circulated in the lower atmosphere as carbon dioxide.

During photosynthesis, this carbon dioxide is absorbed by plants which then maintain during their lifetime an approximately constant proportion of carbon-14.

Man and other animals live on these plants and themselves incorporate this natural radiocarbon.

After death, the plant or animal ceases to gain any more carbon-14, but that already in the wood, charcoal, bones, or shell left behind is decaying at a precisely known rate. Measurement of the proportion of carbon-14 remaining then allows us to calculate how much time has elapsed since the plant or animal lived on Earth.

Above: Diagram of the natural cycle of radiocarbon from its formation in the upper atmosphere to its final decay.

Part of the radiocarbon dating apparatus in the British Museum Research Laboratory. Organic matter from archaeological sites must be converted into a pure chemical compound for this technique to be carried out; although simple in theory, this requires very sophisticated equipment for reliable results.

sample, the date of these rings could be determined, and the sequence of rings in the older sample could be labelled. This process was repeated with older samples of wood until it could be used to date the wood found on an archaeological site.

It is usual to compare tree rings with wood all growing in the same geographical area, as climatic variations over a great distance make the matching of rings rather uncertain. By means of this method, dating has been carried back over some 8000 years. A recent important use of this *dendrochronology* is in the checking or calibration of other dating methods which do not rely on such a simple principle.

Chemical changes

Various relatively simple chemical methods have been used to acquire information about age, particularly of bones and teeth. These materials consist of two main parts: an organic compound, the protein collagen; and a mineral called hydroxyapatite which is a calcium phosphate. On burial, two things happen to these materials. The collagen breaks down and the soluble products are washed away, with the result that the quantity of nitrogen in the bone decreases with time. Simultaneously the mineral part of the bone picks up small traces of certain elements from the groundwater, in particular fluorine and uranium. The bone or tooth can therefore be analysed and the smaller the amount of nitrogen found, and the larger the amounts of fluorine and uranium, the older the specimen must be. It must be remembered, of course, that the rate at which these changes occur depends strongly on the conditions of burial, and so this method cannot be used for making reliable comparisons between different sites. Fluorine dating was first proposed by Middleton in 1844, and acquired fame in the spectacular revelation of the Piltdown forgery in 1953 when allegedly Pleistocene jaw and cranium fossils were shown by analysis to be comparatively modern.

Radioactive dating

Among the most important dating techniques in use at present are those using radioactive substances. These techniques are all based on the same set of principles. It is well known that a number of isotopes of the chemical elements are radioactive – that is they decay spontaneously at a known (but differing) rate, accompanied by the emission of radiation of various kinds. Each different radioactive isotope is characterized by a half life; that is, the time taken for half of the atoms in any given sample to decay. After the same time has elapsed again, half of the remaining atoms decay, and so on. So if one knows the half life of the isotope and the proportion of the atoms of the isotope which have decayed, it is possible to work out how much time has elapsed since the decay process started.

The best known and most generally useful radioactive dating method used on archaeological materials is radiocarbon dating. In the upper atmosphere cosmic ray neutrons react with nitrogen atoms to form atoms of the radioisotope carbon-14. In due course this reaches the lower atmosphere and in the form of carbon dioxide is absorbed by plant life and, through the food chain, living organisms in general. During its life therefore a plant or animal acquires an approximately constant proportion of carbon-14 in the carbon compounds of which it is composed. After death, however, no further carbon-14 is added, and that already present decays with a half-life of about 5570 years. If, therefore, a suitable piece of organic material – for example charcoal, bone, or antler – is selected from an excavation and the proportion of carbon-14 in the remaining carbon is determined, it is possible to calculate the number of years which have elapsed since the organism concerned died. There are, of course, complications: the amount of carbon-14 is so small that the apparatus used has to be very sophisticated, and it is not quite true that the amount of carbon-14 reaching Earth has been constant. But the method has proved of inestimable value for dating archaeological sites and individual finds up to about 60,000 years old, and has been much used for obtaining late glacial and postglacial dates and checking early historical chronologies.

Another very important isotope dating technique, and one which can be used for periods well before the arrival of man, is potassium argon dating. Many rocks contain potassium, which itself contains a minute quantity of the radioisotope potassium-40. This decays to form argon, an inert gas which may remain trapped in the rock. If the rock becomes melted, for example as a result of volcanic activity, this argon is lost and the 'clock' is reset. If, therefore, some fresh volcanic rock is deposited on an archaeological site it initially contains no argon, but this is accumulated with the passage of time as the potassium-40 decays. If a sample of such rock is taken from a context whose date is required, and the amount of potassium and argon in the rock is determined, the age of the sample can be calculated. The half-life of potassium-40 is long, over a thousand million years, and this method is most useful in the early part of the palaeolithic where carbon-14 dating cannot be used.

There are other ways in which radioactive isotopes can provide information about dates. The processes of radioactive decay can produce actual damage in the minerals in which they occur. For example the fission (splitting) of certain heavy atoms like uranium-235 gives rise to heavy particles which leave tracks in the mineral. These can be developed by polishing the mineral and treating it with a suitable acid, when the tracks are visible under an ordinary microscope. It can be seen that for a given mineral containing a known amount of the radioisotope, the number of tracks will be proportional to the age of the sample. Most rocks will show many of these tracks, but if the rock becomes heated and is deposited in an archaeological context, the old tracks are destroyed and it can be used for dating purposes. Here once more, volcanic materials on sites can be used for this *fission track dating* as it is called; and the natural volcanic glass obsidian, which was used in antiquity for the manufacture of artefacts, has proved valuable in this respect.

A pottery clock

One final dating method which is relatively new but has proved to be exceptionally useful is *thermoluminescence dating*. This also makes use of the damage caused to minerals by the decay of radioactive isotopes, but here the damage is measured by heating a small quantity of the material when it will emit a minute amount of light. This light represents the energy stored in the minerals as a result of radioactive decay, and if it is accurately measured, together with the amount of radioactivity to which the sample has been exposed and its sensitivity to radiation, it is possible to calculate the period which has elapsed since the sample was last heated. This method is particularly valuable for dating materials which have previously been heated by the agency of man; by far its greatest use has been in the dating of prehistoric pottery, but it has also been used for dating clay from hearths, and various types of stone which have become burned.

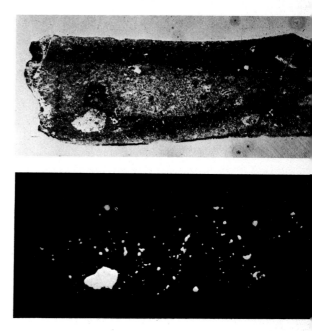

Above: Two photographs of a thin section of Roman pottery, one taken under normal light (top) and the other taken by the thermoluminescent glow produced by irradiation and heating. The glowing areas correspond to the position of mineral grains, seen in the daylight photograph. Below: Diagram showing how thermoluminescence can be used in dating.

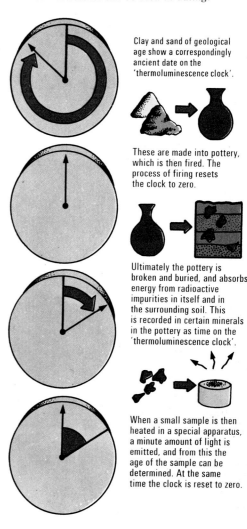

Clay and sand of geological age show a correspondingly ancient date on the 'thermoluminescence clock'.

These are made into pottery, which is then fired. The process of firing resets the clock to zero.

Ultimately the pottery is broken and buried, and absorbs energy from radioactive impurities in itself and in the surrounding soil. This is recorded in certain minerals in the pottery as time on the 'thermoluminescence clock'.

When a small sample is then heated in a special apparatus, a minute amount of light is emitted, and from this the age of the sample can be determined. At the same time the clock is reset to zero.

The Process of Evolution

A group of Lapps, in their traditional clothing, herd their reindeer. The Lapps have a cultural tradition quite different from that of the other Scandinavians, and blood analysis shows that they are also biologically quite distinct.

Man has been shaped by the same general evolutionary laws which influenced the earliest forms of life, and the giant dinosaurs which lived so many million years before him. The physical changes seen in the fossil primates and hominids were controlled by the basic rules of evolution, and Man's cultural development too was closely related to the increase in his brain size and changes to his behaviour. More than anything else, of course, it was this great piece of evolutionary luck – the ability to evolve a far more complex brain – that has enabled Man to step away from all other animals. It has also enabled him to become more independent of the natural world, and gradually to exploit it as no other species has done in the past or is likely to do in the future.

The varieties of hominids, like those of other evolutionary lines in the animal world, have changed through time, and we see the same sorts of patterns of change. Genera and species gave rise to others, and while some varieties were dead-end lines and eventually became extinct, others evolved into more advanced peoples, who then moved off into different and varyingly isolated geographical regions, so that further small-scale evolution usually occurred in relation to the new circumstances. This is the reason for the differences one can see even in the closely related human races today.

In studying the cultures of the past, rather than the physical differences seen in these earlier peoples, it is important to keep in mind how complex the links have been between biological man on one hand, and cultural man on the other. We know from the fossil evidence that between about five and one million years ago, primitive man began to move upright. This left his hands free for other work, especially tool-making. This sort of cultural development probably called for better co-operation between individuals in hunting, and better communication by language. These extra needs may then in turn have influenced the physical evolution of larger brains to cope with them. So there has been a cycle of biological and cultural events all influencing human evolution and this is still a continuing process today, although very different in kind from the days of the earliest farmers or the earliest tool-making peoples.

The biology behind evolution

Man reproduces his species in the same way as most other mammals, and the way he inherits characteristics is also the same. Much of the physical variation we see in fossil, historic, and recent peoples can be related to basic mechanisms of inheritance.

Most of us are aware of certain similarities between the different generations of our own family. We may notice similarities to our parents, or even grandparents, in the colour of our hair or eyes, how tall we are, or even such finer details as nose or ear shape. At the core of this handing on of similarities is the gene, a biochemical unit of great importance. It is combinations of these units which provide a blueprint for the correct development and growth of each individual. This process is only changed if an error occurs in the basic chemistry of a gene or if the environment influences the greater or lesser expression of the variety of genes.

Not all genes have equal importance or express themselves to the same degree. Some are called *dominant*, because the characteristics they control reveal themselves clearly. The action of these dominant genes may mask the influence of other, *recessive* ones. These 'silent' genes usually express themselves far less commonly. But most genes really fall between these two extremes, and seem often to work together to influence the expression of other genes. Each individual is, then, the result of the total action of all these genes, some strongly dominant, others recessive, and probably most with varying degrees of influence on others. Superimposed on this genetic level is the 'environment', which can modify our development and health before birth as well as in later years. The woman who smokes during pregnancy may stunt the growth potential of the baby within her. Similarly a scarcity of food after birth may prevent a child from reaching the height it is genetically programmed to achieve. The term *phenotype* refers to the individual as he turns out to be – the addition of environment to genes – and it is on this state that evolution acts.

Evolution is really the result of many success stories. It is concerned with surviving, first to the point of being born, then into adult years. In particular, it is the individuals who produce healthy children who determine the future of their species. As Charles Darwin recognized many years ago, it is this ability to survive on the part of a certain proportion of a population which leads to small-scale evolutionary changes over the generations.

A group of people has a quantity of varying genes available to them to pass on – through their children – to other generations. This is known as a gene pool. If every member of the community does not reproduce the same number of children, and they are not equally able to survive and have further children, the composition of the gene pool will change. Some genes seem to confer better health or in other respects a greater ability to live into adulthood and reproduce; more people possessing them will survive to pass them on so the gene pool will have a higher proportion of them. In the same way some genes are likely to be gradually eliminated from the gene pool – depending on

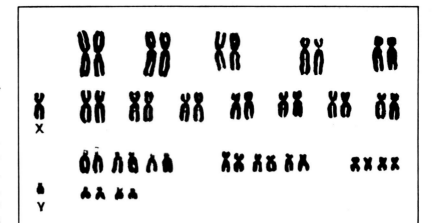

The 46 chromosomes of a human male. They have been sorted into 22 pairs; the two left over are strikingly dissimilar. These are the X and Y chromosomes. It is the Y that determines the male; a female has two X chromosomes.

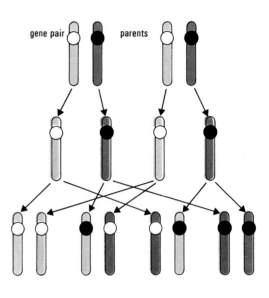

gene pair   parents

The possible combinations of genes in the offspring of parents each with a dominant gene for a yellow characteristic and a recessive gene for a green characteristic. Each parent is yellow. Each can pass on one gene to its offspring; these may be yellow, yellow but with a recessive green gene, or purely green. Recessive genes can be handed down unnoticed for many generations, only appearing in the absence of a more dominant gene.

GENES

The characteristics of all living things are inherited from their parents. This takes place mainly by the action of genes – biochemical units which provide the blueprints for and instigate growth and development. The combination of genes in an individual is unique, since the processes of reproduction involve a combination of some genes from both parents which is different in each individual.

Genes are carried on chromosomes – threadlike structures in the nucleus of the cell. Almost every cell in any particular species has the same number of chromosomes. Take, for example, a human; he or she has 46. Half of these come from the father and half from the mother, and each of these sets of 23 contains a complete collection of genes. Some of these are more dominant than others – for example, a child that inherits one gene for brown eyes and one for blue eyes will always have brown eyes, since that is the dominant gene. But on the whole the genes interact in a more complex way to determine the development of the individual.

A parent has two gènes for every characteristic; and it passes on only one of each. This *reduction* in the genes takes place in the formation of the sex cells, the sperm and the egg. They are formed by a process called meiosis in which a cell with the full number of chromosomes – 46 in the human – divides into cells with only half the chromosomes. During meiosis the original chromosomes sort themselves into pairs bearing genes for the same things, and these pairs exchange pieces. As a result chromosomes are formed which contain exactly the same genes, but in different combinations. So a man does not hand on the chromosome set inherited from his father or that inherited from his mother, but a unique mixture of the two. This process allows the widest possible variations and combinations of genes, and hence of individuals; and it is essential to evolution.

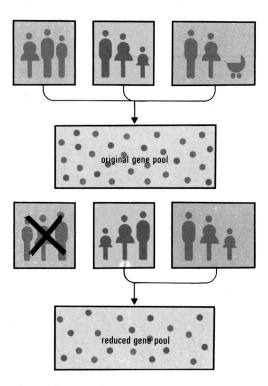

How the death or emigration of a family can affect a community's gene pool by reducing the variety of available genes. The greatest effect is seen in small communities where the available gene pool is comparatively limited.

what selective factors are influencing the success of the different genes. Variations in a community will change depending on these selective pressures on the gene pool. But evolution also demands that there should be a continuous supply of additional variations in the community.

Mutations

The most important source of additional variations is *mutation* – a change at a chemical, general structural, or numerical level affecting our development and survival. Mutations are of two major kinds. First are changes to chromosomes – the basic structures within the cell nucleus, along which the genes are placed. Such chromosome changes occur during cell division, and may result in parts of chromosomes (and thus blocks of genes) being lost, or moved into different positions. Some abnormal states of development in man, including some forms of mental defect, are the result of having more than the usual number of 46 chromosomes. Perhaps most important in our evolution has been chromosome mutation resulting in modifications to the lineal arrangement of genes along the chromosomes, so that the coding of growth might be changed.

The second kind of mutation is that of actual genes. This results in some degree of biochemical change, simple or more

This much simplified diagram shows the different kinds of structural alterations which can occur in the chromosomes.

complex. At times, even the modification of a fairly basic chemical factor may have profound effects. An example is the gene for an abnormal condition known as phenylketonuria. This is a recessive gene, but if both parents carry it by mutation or inheritance from past generations, their children can receive the gene from both of them and thus develop the disease. As a result they are literally poisoned by the inadequacy of their own body chemistry. They may also suffer brain damage and some modification of growth, and even their hair colour may be affected. This is a startling example of the effect one gene can have, but many gene changes must result in only slight changes of normal characteristics, and these, of course, are the ones which provide additional variations which play such an important part in evolution. Fortunately some mutations have occurred more frequently than others in the course of human evolution, and through this great genetic luck our species has emerged with its specialized way of moving, sensitive hands, and uniquely complex brain.

Evolution at population level

Evolution can be studied at different biological levels: modifications to the minute chemistry of genes, the expression of these in the individual as a whole, how well they permit survival in the environments in

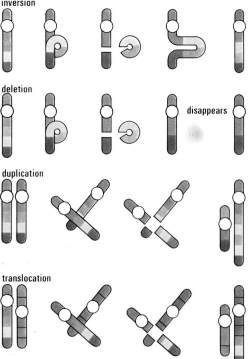

inversion

deletion

disappears

duplication

translocation

which people must live, and finally how each individual is placed in relation to others in this world as a whole, and particularly in relation to his own more local community. Evolution is all this added together and seen through the perspective given by time.

In comparing distinctive varieties of fossil man, spanning many thousands of years, we are seeing and can appreciate even in one individual the result of considerable genetic differences between the different forms. Even a single skull from each species may be enough to demonstrate the marked genetic gaps between each particular species and the others, although not showing the variation present in each community as a whole. But it will not tell us why the divergence between the groups occurred. Only by the careful study of small-scale changes in recent peoples, and in other mammals, can we hope to understand the reasons for particular changes more fully. So the study of *micro-evolution* – smaller changes in peoples over more limited time – is very important in giving us a proper understanding of human evolution.

One of the best examples of how small-scale genetic changes can occur in man in various regions is concerned with an abnormal blood condition. Haemoglobin is a pigment which makes up over 90 per cent of the protein content of a human red blood cell. An abnormal form, haemoglobin S, can be inherited from one parent or both (giving gene pairs NS or SS – with single or double quantities of the abnormal haemoglobin). **Someone with SS genes has** *sickle-cell* **anaemia (named from the cell shape), with little chance of growing up or of having children. Normally** we would expect this elimination of S genes to alter the frequency of the genes for the normal and abnormal haemoglobins over the generations, as more of the genes causing severe anaemia were lost. But in this particular example the story is rather more complex. A study of the condition showed that in certain regions where the S haemoglobin occurred, especially in Africa, the frequency seemed to be surprisingly high. Investigations brought to light three significant facts. First, the communities who were most affected lived in certain areas where malaria was common. Then it appeared

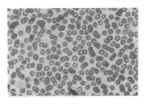

The microscopic appearance of normal red blood cells (above) compared with those from an individual with sickle-cell anaemia, both magnified 208 times.

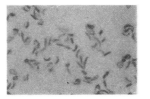

The distribution of malaria in comparison with the severe anaemia associated with the abnormal haemoglobin S. The anaemic condition is often referred to as sickle-cell anaemia because the red corpuscles may be distorted into a sickle shape. Other forms of abnormal haemoglobin are found in the Mediterranean area and Asia and seem also to be associated with malaria resistance.

that in these groups, the individuals without the abnormal haemoglobin died from malaria at about the same rate as the SS individuals died from severe anaemia. But the people who inherited a gene for normal haemoglobin (N) plus an abnormal gene S from their parents seemed better able to survive the invasion of the microscopic malaria parasites into their blood.

So a study of this abnormal haemoglobin tells us something about evolution in miniature. First of all, the two genes N and S vary in frequency in different groups, and S is absent in some. The S gene weakens the health of individuals who carry it, and this factor on its own would probably lead to the survival of more individuals with only N genes. But in the malarial areas the NS people have some health advantage over both NN and SS members of the community. So a state called a *balanced polymorphism* is established – that is, neither gene is noticeably reduced because NS people are biologically most 'successful'. Such a condition is only stable if the selective pressure, this preference for the 'middle man' in areas of malaria, continues to operate. As a result of the world-wide anti-malarial campaigns of recent years, the gene has lost what value it had to man, so that gene pool changes will occur and the S gene will become increasingly rare.

This condition is, of course, by no means the only variation seen in the blood, or even of the red-blooded cells. It is generally known that individuals may have different blood groups, and that hospitals make use of this information in blood transfusion work. The information used by hospitals is usually in relation to two of these groups, the ABO and Rhesus systems, but in fact many more blood groups

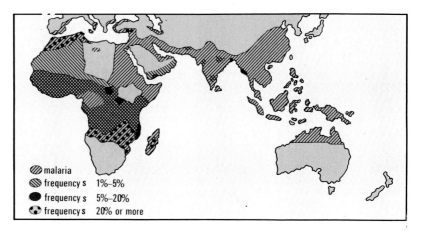

malaria
frequency s 1%–5%
frequency s 5%–20%
frequency s 20% or more

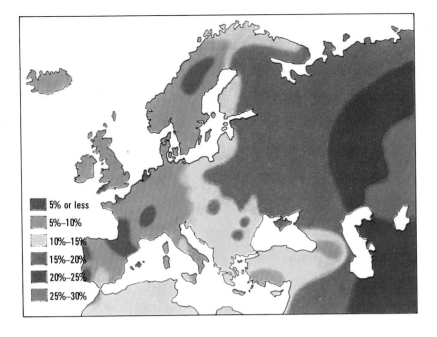

Frequencies of the gene B of the AB0 blood group system in Europe. There is a gradual gradation in frequency from east to west. Such frequency differences also occur in other areas, and for other blood group data.

5% or less
5%–10%
10%–15%
15%–20%
20%–25%
25%–30%

are known, and show differences in the proportions of the genes present from one community to another.

From the point of view of micro-evolution, over twelve separate blood group systems show significant regional differences. In the case of the ABO group, there are some noticeable world differences, and even within regions such as Europe changes in frequency can occur. These clines, or gradual regional changes, may well be related to the history of peoples (especially over the past 30,000 years or so), their movement into different areas, and their success in establishing new communities or mixing with ones already there. It can also reflect the ability of cultural groups to maintain their independence over long periods of time. The earliest American Indians who had crossed the Bering Straits from Asia into the New World were restricted in numbers and therefore in the genes they carried in themselves. The result was a genetic bottleneck, which was later reflected in their descendants, for in a number of blood group characteristics they are unlike the Asians who are their closest living relatives. This so-called *founder effect* must have happened many times during evolution, especially with small nomadic bands.

Returning to Europe, the Basques of northern Spain still show differences in their blood group factors from those of the nearby Spanish and southern French groups. This emphasizes that they have remained apart from their neighbours – not only in terms of language and other cultural characteristics but to some extent biologically as well. The Lapps of northern Scandinavia, too, show not only a long tradition of cultural separation but also evidence of biological distinctiveness. This latter variation includes frequency differences for the blood, where the genetics has been well studied, and also physical differences where many genes are involved together in producing the structures and variation – often referred to as *multifactorial inheritance*. The facial features of the Lapps are distinctive, and there is even evidence that a form of hip dislocation – again controlled by many genes – is especially common among them. This is probably an example of another evolutionary phenomenon which has been termed *relaxed selection*.

During the palaeolithic cultural phase, human groups must have been largely nomadic or semi-nomadic, small in size, and dependent on their immediate surroundings for food. Inherited defects such as colour blindness may have been a considerable disadvantage to hunters and collectors, who had to distinguish partly camouflaged animals in the bush or identify by colour as well as shape which berries were edible. But with the development of farming and the specialization of activities, the colour blind person and others with similar mild defects would have been increasingly protected and could have turned their hands to trades where colour identification was not important. So what may have been critical to palaeolithic hunters, perhaps having repercussions on the survival of themselves and their families, became transformed into a variation of little importance. The evolutionary situation was in other words 'relaxed'; the genes survived and were handed on to other generations.

These basic evolutionary and genetic factors have been responsible for the evolution of man and the development of his increasingly complex cultures, so that he has attained a unique position in the animal kingdom. His safe journey into the distant future will be a special challenge to the complex brain he has already evolved, but further biological and social adaptations – perhaps in part engineered by scientists – may still be necessary.

How Primates Evolved

A reconstruction of Plesiadapis, a prosimian of the Palaeocene epoch living about 65 million years ago. It was very like a squirrel in appearance, and since it had claws rather than nails it probably moved like a squirrel. It had similarly enlarged front teeth, so its whole way of life must have been very squirrel-like.

Primates is the zoological order to which man belongs. As such it is specially interesting to us, since a thorough understanding of the interrelationships and evolution of primate species will lead to a better understanding of these things in relation to man.

All primates share certain characteristics which have contributed to their evolutionary success. Their major characteristic is their *lack of specialization* to any one environment or ecology. They are mostly adapted to life in trees (arboreal), but many are terrestrial, living far from trees, and even the arboreal species show an enormous variation in the ways in which they move about and their adaptation to their environment. There are many difficulties in defining a group of animals whose main characteristic is lack of specialization. Some of these become apparent when we examine the list of primate characteristics.

The ancestral mammals of the Cretaceous period were small, clawed, arboreal creatures. Most later groups of mammals became different from this ancestral type by developing specializations – that is, characteristics that fitted the animals very well for one thing but which, in so doing, restricted their activities to one way of life. This has produced the anteaters, the moles, whales, horses, and so on, each characterized by a very distinctive mode of life. Not so the primates; most of them have retained or no more than modified the ancestral features of the Cretaceous mammals. This means that early primates in the fossil record are hard to distinguish from their earlier, non-primate, ancestors; and the same is true all the way up to the present day, each stage being difficult to distinguish from the ones before and after it. The evolutionary trends that can be followed are of a very general kind, such as overall increase in size, relative and absolute increase in size of the brain, reduction in snout size and number of teeth, and so on.

The primates in space and time

The order Primates is divided into two, the Prosimii and the Anthropoidea. The prosimians (*pro*, before; *simia*, monkey) are mostly nocturnal animals, lemurs, lorises, and tarsiers, living today in the tropical regions of Africa and Asia. They

CHARACTERISTICS OF THE PRIMATES
Retention of five fingers on hands and feet.
Either hands or feet, or both, are capable of grasping.
Claws replaced by flattened nails.
Retention of clavicle.
Teeth have simple cusp pattern.
Three and sometimes four kinds of teeth are developed (e.g. incisors, canines, premolars, and molars).
Face or snout projection reduced.
Orbits enclosed by bone.
Eyes and optic centre of brain greatly developed.
Nose and olfactory centre of brain reduced.
Enlargement of brain.
Elaboration of cerebral cortex.
Long and elaborate gestation period (4–9 months).
Increased infant dependency time.
Highly gregarious with complex social structures.

*Four types of primate loco-
motion – vertical clinging
and leaping (a tarsier), a
baboon walking quad-
rupedally, a brachiating
gibbon, and a knuckle-
walking chimpanzee.*

product of the competition – one of the
ways in which the apes were forced to
adapt in order to survive the spread of the
more successful monkeys.

The prosimians

The living prosimians are mostly small
insect-eating animals. Once widespread
throughout the world, today they are confined to relic areas where they are safe
from competition with other primates.
They have been bypassed by the main
stream of primate evolution. With a few
exceptions they are confined to the tropical
forests of Africa and Asia. The lemurs
(family Lemuridae) are all found in just
one place, the island of Madagascar, where
they have been isolated for at least the last
30 million years. Their survival there has
almost certainly been dependent on the
absence of higher primates (Anthropoidea), which had not evolved at the time
when Madagascar became separated from
the African continent. The other two main
groups of prosimians, the lorises and the
tarsiers, are not geographically isolated in
the way the lemurs are; but they are
ecologically isolated – their way of life
separates them from most other animals
that might be likely to compete with them.
Of particular importance here is that they
live in trees, are active at night, and are
able to grip and move about on extremely
small branches because of their grasping
hands and feet.

There are just over 20 species of lemur
living in Madagascar. A few are quite

are very common in the fossil record in
deposits ranging from 40 to 60 million
years old, but not so much is known of
their more recent history. The anthro-
poids, or higher primates, also mostly live
in tropical places, although they range
more widely into warm temperate regions;
almost all are active during the day. They
are a flourishing group today, consisting of
the Old World and New World monkeys,
the apes of Africa and Asia, and man. They
did not appear in the fossil record until 30
to 40 million years ago, but spread rapidly
and displaced the prosimians from most of
their habitats.

The prosimians today are in fact relic
populations that have survived the competition from the higher primates by vir-
tue of being small nocturnal, insect-eating
animals. Similarly the apes, from which
man is an offshoot, were the first of the
groups of the higher primates to diversify
and replace the prosimians; they have now
been displaced by the monkeys over most
of their range. In a sense, therefore, man
comes from one of the less successful primate groups, a group that flourished 10 to
30 million years ago but many species of
which are now threatened with extinction.
Man took his origin some time towards the
end of this period and might in fact be a

CLASSIFICATION OF THE PRIMATES	
Sub-order Prosimii	
Superfamily Lemuroidea	
Family Lemuridae	
Adapidae	Lemurs
Superfamily Lorisoidea	
Family Lorisidae	Lorises
Galagidae	Bushbabies
Superfamily Tarsioidea	
Family Tarsiidae	
Anaptomorphidae	Tarsiers
Sub-order Anthropoidea	
Superfamily Ceboidea	
Family Cebidae	New World
Callithricidae	monkeys
Superfamily Cercopithecoidea	
Family Cercopithecidae	Old World
Colobidae	monkeys
Superfamily Hominoidea	
Family Hominidae	Man
Pongidae	Apes
Hylobatidae	Gibbons

large and have a distinctive method of locomotion, called vertical clinging and leaping, by which they jump from one vertical branch or tree trunk to another with their long and very powerful legs. These animals are all diurnal – active during the day – and they live in small social groups. The other main group of lemurs is more varied. Some are nocturnal, but others are diurnal and live in quite complex social groups. They have longer arms than the vertical clingers and leapers and use all four limbs in running and leaping.

The Lorisidae are divided into the galagos of Africa and lorises of Asia and Africa. They differ from the lemurs in the way they move about and in certain features of the skull and teeth, but the two groups share a number of characteristics, including the possession of comb-like front teeth used for grooming. The galagos are nocturnal and move by vertical clinging and leaping like some of the lemurs. The lorises are also nocturnal, but the way they travel around could not be more different, for they move very slowly, one leg at a time, and with their large grasping hands and feet they are almost impossible to dislodge from the branch they are climbing on.

The final group of prosimians, the tarsiers, are again nocturnal. They also move by vertical clinging and leaping, but they are strikingly different from the lemurs and lorises in their skull anatomy, having large eyes and a dry furry nose as opposed to the moist nose of lemurs. They have fewer teeth than the lemurs and lorises and no dental comb.

Old and New World monkeys

There are three major divisions of the Anthropoidea, the New World monkeys (Ceboidea), the Old World monkeys (Cercopithecoidea), and the apes and man (Hominoidea). The New World monkeys are an extraordinarily diverse group. They have been isolated in the Americas for a very long time, so that despite their name they are not at all closely related to the monkeys of the Old World. Like the lemurs of Madagascar, the New World monkeys evolved in isolation without any competition from other primates, but because they had a longer time and a greater variety of habitats than the lemurs they have evolved further and done so in a

way that parallels in some respects the Old World monkeys.

The Old World monkeys are divided into two rather distinct groups – the leaf-eating colobus monkeys (Colobidae) and the omnivorous guenons, baboons, and macaques (Cercopithecidae). All are diurnal, and they are remarkably similar in most aspects of their skulls and teeth. The colobus monkeys spend most of their time in trees, but some of the cercopithecines are largely ground-living. The latter are the most abundant and the best known. The baboons and macaques have developed the most complex social structures, but all of the monkeys have more elaborate social organizations than the prosimians.

Apes, gibbons, and men

The final division of the Anthropoidea consists of the group that includes the great apes, gibbons, and man. The gibbons, or lesser apes, are today found only in the forests of South-east Asia. They are characterized by the way they move – swinging by their arms which are extremely long and powerful. But they also have long legs, and, after man, they are more often seen walking bipedally (on two feet) than any other primate. They are strictly fruit-eating in their diet.

There is some evidence that the gibbons' skulls and teeth are the most primitive of all the higher primates, so they provide a good model for arriving at the probable diet of primitive primates. But although the gibbons may be primitive in respect of their skulls and teeth, they are definitely very advanced or specialized in respect of their limb bones, so it is not possible to

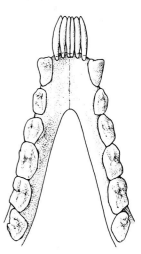

The lower jaw of a lemur, showing the characteristic arrangement of the front teeth to form a comb used in grooming.

A ring-tailed lemur and its young. Lemurs are found only in Madagascar where they have been able to flourish in the absence of competition from higher primates. The ring-tailed lemur is the size of a cat; it lives in groups of a dozen or so and is active in daytime.

speak of them as altogether primitive or altogether specialized.

The great apes include the chimpanzees, gorillas, and orang-utans. They are all large, being the biggest of the non-human primates, and as a result of their size they are less agile than most other primates. The orang-utan is a solitary arboreal animal that moves slowly in the trees, usually keeping a grip on the branches with at least three of its hands and feet while moving the fourth to a new position. The orang-utan is found today in Borneo and Sumatra; but the other two, the chimp-

anzee and gorilla, are found only in Africa. Although both are forest-living animals, they spend much of their time on the ground, so they cannot be said to be arboreal. The chimpanzee climbs trees to eat the fruits growing on them, but to move from one tree to the next it almost invariably comes to the ground and walks quadrupedally, taking its weight on its arms through the knuckles of the hands. This odd position is known as knuckle-walking. The gorilla walks in exactly the same way but spends nearly all of its time on the ground. As it is so big and heavy it

The distribution of the primates today. Some of the groups are represented here by single species; the New World monkeys are represented by the spider monkey and woolly monkey, and by a marmoset. All the apes are illustrated – the chimpanzee and gorilla in Africa, and gibbon and orang-utan in Asia. The baboon and Barbary 'ape' in Africa and the langur in India and Asia represent the enormous variety of cercopithecoid monkeys on the two continents, and prosimian distribution is shown by two lemurs from Madagascar and a tarsier and a loris from Asia.

44

has to be very careful if it does climb a tree. It eats leaves and plants rather than fruit.

Early primates

Primates are first known definitely from the fossil record at the beginning of the Tertiary Era, that is about 70 million years ago. At this time many of the other groups of mammals had also evolved into distinct categories – for instance the carnivores and insectivores – and there are many other fossil forms which have no modern descendants or whose relationships with later forms is not clear. The early primates are picked out from these mainly on the basis of their teeth and skull characters, for in most other respects they were indistinguishable from the insect-eating animals who were their ancestors. They were small, tree-living creatures with claws rather than nails, a projecting snout, and poorly developed binocular vision. Compared with the list of primate characteristics given earlier they were hardly primate at all, but this is to be expected in a form that has only just branched off and begun evolving towards the living primates.

45

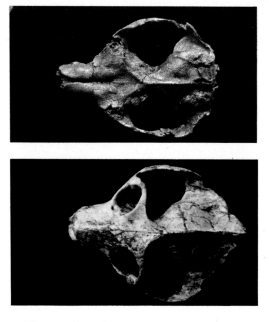

The earliest known primate is *Purgatorius*, known from the late Cretaceous. The structure of its molar teeth is very like that of later primates, but it is also so like the contemporary condylarths (primitive ungulates) that its identification as a primate is still doubtful.

In the Palaeocene several families of primate are known. All of these had peculiarities of the teeth in one form or another – for example rodent-like incisors or large serrated premolars – but the structure of their molar teeth is like that of later primates. The best known of the Palaeocene primates is *Plesiadapis*, fossils of which have been found in both Europe and North America. It was a squirrel-like animal with claws instead of nails and it had not developed the grasping hands and feet typical of later primates. Although it was almost certainly arboreal it must have moved about the trees in a squirrel-like way, rather than in the manner of later primates – it ran along the tops of branches rather than hanging from them. *Plesiadapis* had a long snout; its eyes were placed on the side of its head and were unprotected by a post-orbital bar (see above), so that clearly it relied more on smell than on sight. The only really characteristic primate feature of *Plesiadapis* and other Palaeocene species is their flat low-crowned cheek teeth with low rounded cusps. These are so similar to later Eocene primates that on this basis alone these Palaeocene forms are included among the primates.

Eocene prosimians

By the Eocene things were very different. The Palaeocene forms had apparently disappeared without trace, and in their place appeared a new and widespread kind of primate. These new forms were clearly primate although they were still prosimian in grade. The best known of these are *Adapis* and *Notharctus*, the European and American forms respectively of the family Adapidae. They still had some primitive features not found in living prosimians, among them four sets of premolars and vertically placed incisors and canines which did not yet form a comb. On the other hand they had developed many of the primate characteristics – for instance reduced snouts; forward-pointing eyes protected from behind by a bar of bone and providing overlapping fields for stereoscopic vision; relatively larger brain size; nails instead of claws; and grasping hands and feet.

Although the adapids are well known from many fossil remains, we still do not know their relationships with earlier and later primates. *Adapis* is so similar to living lemurs that we can assume they are in some way related, but there is no trace of any intermediate forms between the Eocene and the present, a period of some 40 million years. In the Miocene of Africa there are some fossil prosimians from deposits about 20 million years old, but these are clearly linked with the bush babies or galagos, and with the lorises; it is impossible to connect them with the Eocene forms.

The adapids were tree-living animals with a very similar way of life to that of the more generalized lemurs of Madagascar today. They were probably not vertical clingers and leapers, as has sometimes been claimed, but four-footed climbers and scramblers through the branches. They were probably largely fruit-eaters, with leaf shoots in season and insects and birds' eggs added when the opportunity arose.

As well as the adapids the ancestors of the tarsiers were flourishing during the Eocene. Some of these were quite primitive, and some developed specializations of the teeth, but the most interesting of the European tarsiers, to which *Necrolemur* belonged, remained relatively unspecialized. A fourth group, the Omomyidae, was very widely spread through North America, and in many respects they are the most generalized of the Eocene prosimians and the ones that could have been the ancestors of the higher primates.

After the end of the Eocene the number of prosimian primates in the fossil record began to decline. It is likely that as the lemurs and lorises developed their distinctive specializations of teeth and limb bones (for example the tooth comb, and vertical clinging and leaping in the limb bones) they became side-tracked from the mainstream of primate evolution and were not able to compete with the higher primates, the monkeys and apes, as they evolved. The lemurs have survived until today only through being isolated on Madagascar without competition. There they have evolved into a great many different forms over the last few million years. The larger species of Madagascar lemur became extinct very recently, in the case of *Megaladapis* no more than 300 years ago. The evidence suggests that these animals became extinct as a result of man's arrival on the island, and today only the smaller and often nocturnal forms still survive.

Early anthropoids
The earliest higher primates known from the fossil record are from the Oligocene period in Egypt. The deposits of the Fayum region date from between 30 and 40 million years ago. The primates found there are definitely related to the Old World primates, and it must be presumed that the Old World and New World primates had evolved separately from different prosimian stocks some time before this, probably towards the end of the Eocene. There are fossils of at least two groups of higher primate at the Fayum. One group, the Parapithecidae, left no descendants. The other main group includes *Oligopithecus, Aegyptopithecus,* and *Propliopithecus,* and these are almost certainly the ancestors of all the living Old World monkeys, apes, and man.

Oligopithecus is known from only a single specimen, but it is important because it is one of the few intermediate fossils in the primate record. Although in most cases we simply do not know how groups of fossil primate originated, nor what they evolved into, *Oligopithecus* is undoubtedly part prosimian and part

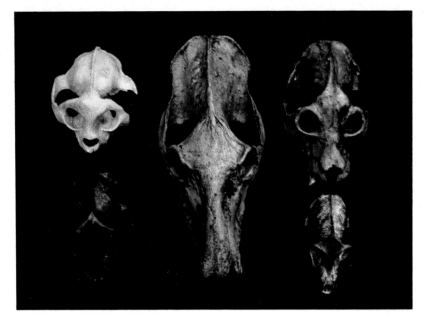

anthropoid. The numbers of teeth and their arrangement clearly link it with the higher primates to come, while the structure of the molar teeth is similar to that of the omomyid prosimians of the Eocene period. *Oligopithecus* was found only in the lower fossil wood zone of the Fayum at quarry E; it lived some 35 to 40 million years ago.

Next in time from the Fayum is *Propliopithecus*. This is sufficiently similar to *Oligopithecus* for there to be little doubt that they were closely related, but it is considerably later, and more advanced in its tooth structure. Similarly *Aegyptopithecus* is later still and even more advanced than *Propliopithecus*, and was probably descended from it. Both look like miniature apes, but while *Propliopithecus* is still quite generalized and could have given rise to any or all of the monkeys, gibbons, or great apes, *Aegyptopithecus*

Skulls of some of the widely differing forms of lemur which evolved in Madagascar. Archaeolemur *(left, below) developed monkey-like features, while* Hadropithecus *(left, above) developed specializations of teeth and skull similar to those of the gelada baboon. The giant* Megaladapis *(centre) was the size of a young calf.*

The Oligopithecus *lower jaw. It has the same number and kind of teeth as modern monkeys and man.*

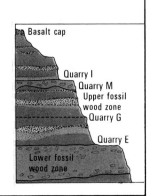

THE FAYUM
The Fayum Depression today is part of the Egyptian desert, but in the Oligocene it was covered with forests and large rivers. Primates are common in the deposits, and were first found at the beginning of this century by a German collector, Richard Markgrof. Many were found in the 1960s when Professor Elwyn Simons of Yale University began an extensive series of excavations. Professor Simons worked mainly in Quarries G and I, from which most of the specimens of *Propliopithecus* and *Aegyptopithecus* came, but his two most significant finds were from other quarries, a skull of *Aegyptopithecus* from Quarry M and a lower jaw of *Oligopithecus* from Quarry E.

Basalt cap

Quarry I
Quarry M
Upper fossil wood zone
Quarry G

Quarry E

Lower fossil wood zone

had apparently already started on the way to the great apes, the chimpanzees, gorillas, and orang-utans, and, at this stage, man as well.

Aegyptopithecus has typically ape-like teeth, with large projecting canine teeth, upper molars with four cusps and lower molars with five, and the cusps are low and rounded. Its lower jaw is deep and heavily built as in apes and quite unlike that of prosimians. Its skull is primitive in some ways, such as the elongated snout, but its eyes are completely closed in behind by bone whereas prosimians have just a single strip of bone, the post-orbital bar, protecting their eye from behind. This development is related to the increased importance of the eyes in higher primates. Its ear region is still primitive. Its limb bones lack the characteristics of later apes, but show that *Aegyptopithecus* must have been a climber, living exclusively in trees and moving about in a way like that of the howler monkeys of today.

The spread of the Oligocene primates of the Fayum marks the earliest radiation of the higher primates. Several species of these early apes lived side by side with several more species of monkey-like primates, the parapithecids. At some time in the Oligocene the Old World monkeys must have become separated and gone their own way, but so far we do not know just when this happened. The same is true of the gibbons.

Miocene apes

Man's earliest ancestors emerged from the Dryopithecinae or Miocene apes – descen-

dants of the apes of the Fayum. *Aegyptopithecus* itself is sometimes included in the Dryopithecinae, but the spread of this group was mainly in the Miocene, between 10 and 25 million years ago. The earliest evidence, dating from the early Miocene, is from East Africa, where the first and biggest expansion of dryopithecines took place. At least six species are known, divided into three groups: *Proconsul*, *Rangwapithecus* and *Limnopithecus*. There is also a single species assigned to the gibbon family and called *Dendropithecus*. These names are all descriptive: *Proconsul* was named after a famous chimpanzee in the London Zoo, Consul, with *pro-* meaning before to show that *Proconsul* was an ancestor of the chimpanzee. *Rangwal* is the name of a Miocene volcano in East Africa, and *Limno-* lake, and *Dendro-* wood or trees, indicate lake-ape and wood-ape.

All the early apes had basically similar teeth to those of the living apes. The single gibbon species, *Dendropithecus*, differed from the dryopithecine apes in having sharp cutting canines and incisors. Gibbons and great apes today have this same difference.

Below: East Africa in the early Miocene period – the surroundings in which the early apes lived. At this time Africa was rather different from the present: the rift valley had only just begun to form and the great areas of highland were not present. There was a number of newly formed and very large volcanoes rising above the forested lowlands, and it is near these that the primate fossils are found. The dryopithecines probably lived both on the slopes of the volcanoes and in the lowlands, and it seems likely that their spread and diversification at this time was linked with the onset of the volcanic activity, which would have produced varying and rapidly changing environments.

We do not yet know what these early dryopithecines evolved into. There is no evidence linking any one of them with one of the living great apes. But the ancestry of the lesser apes or gibbons is clearer. *Dendropithecus* is similar to gibbons in its tooth structure and what is known of its skull, and its limb bones have started to become elongated like those of the living gibbons. It probably did not swing from its

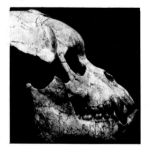

A reconstruction of Proconsul, showing it high up in the trees in its natural environment. It did not walk on the knuckles of its hands, as do living chimpanzees and gorillas, but in most other respects it looks like an ancestral chimpanzee or gorilla – that is, an ancestor of both of these and of man, before they had become separated in the course of evolution. Below is a Proconsul skull.

arms like the gibbons, but its anatomy could certainly have evolved to allow this. So the evidence suggests that *Dendropithecus* was a direct ancestor of the gibbons.

The early apes were quite like monkeys in their habits, but they are recognizable as apes from their teeth and certain features of their skulls. The anatomy of their arms and legs in some ways foreshadows that of their descendants, the living great apes, but was for the most part monkey-like, so the way they moved about in the trees must also have been essentially monkey-like.

Some of the early Miocene apes of about 20 million years ago survived in East Africa until the middle Miocene of Fort Ternan. But most disappeared from the fossil record and were replaced in Europe and Asia by a second wave of

dryopithecines. At least five species of *Dryopithecus* (the oak ape) are known from 14 to 10 million years ago, and these include the first dryopithecine fossil ever found, the lower jaw of *Dryopithecus fontani* found in 1856 in France.

From the same period, but only very rarely from the same deposits, came another group of European Miocene apes called *Pliopithecus*. These have long been thought to be fossil gibbons, and they are similar in many respects to *Dendropithecus* of Africa, but they have a number of characters which are less advanced than those of *Dendropithecus* even though they lived several million years later. So evidently they represent a side branch of the gibbon family rather than a direct ancestor.

Two other primate side branches also lived in the middle Miocene of Europe and Asia. One of these, aptly named *Gigantopithecus,* was the largest primate ever known to have existed – much larger than the gorilla today. Its descent from the dryopithecines can be traced, but it apparently left no descendants itself, unless the abominable snowman exists. The other, called *Oreopithecus,* has no known ancestors or descendants.

Unfortunately we have as yet no evidence linking any one of these early apes with any of the living great apes. It has been suggested that two of the species of *Proconsul, P. africanus* and *P. major,* were direct ancestors of the chimpanzee and gorilla respectively, and *Dryopithecus fontani* was also compared with the chimpanzee, but it is now recognized that there is so much variation in the Miocene apes that it is no longer possible to single out any one of the species as ancestor. It is almost certain that *one* of these early apes was ancestral to some or all of the living great apes, but which one this could be is not known.

Evolution of the monkeys

The monkeys have followed and replaced the apes in evolutionary success. They are first known from fragmentary fossils from the early Miocene, from the same deposits that yielded so many remains of ape fossils. They outnumbered the apes at a later site in Kenya, Maboko Island, about 15 million years ago, but they apparently did not become widespread until much later – 10 to 8 million years ago – when they are

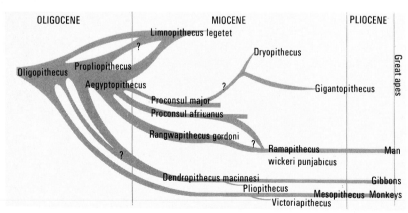

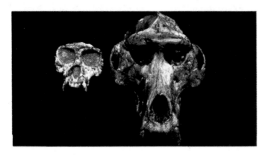

Two fossil monkeys – Mesopithecus (left) which became common in Europe as apes declined in number, and Simopithecus whose large molars and small canines indicate small-object feeding and hand-use – the pattern which is the most plausible model for man's evolution.

found in large numbers in many sites in Europe. The division into colobines and cercopithecines had probably already taken place at Maboko, and it was colobines like *Mesopithecus* that first became common in Europe, significantly at a time when apes were becoming rare in the fossil deposits. Today the colobines are rather less successful than the cercopithecines, and it is probable that there is a trend, which must have begun several million years ago, in favour of the cercopithecines.

A group of fossil cercopithecines named *Simopithecus* was widespread several million years ago, but today they are represented only by the rare gelada baboon of Ethiopia and their place has largely been taken by the true baboons. The interesting thing about them is that they have very large molars and small canines and incisors, as in man, and that like man they use their hands a lot in feeding. It is this use of their hands that has allowed the reduction in size of the incisors and canines, no longer so important in tearing up food, and it is their diet of small objects like seeds that has brought about the increase in size of the molar teeth for crushing and grinding. This pattern of small object feeding with the emphasis on hand-use is the most plausible model for the evolution of man.

The earliest man-like apes

As long ago as the 1930s G. Edward Lewis of Yale University noticed some dryopithecines that looked remarkably manlike. These are called *Ramapithecus*, first described from India but now known

A reconstruction of Oreopithecus, showing it in the swamp environment in which it is nearly always found in fossil form. This ape has no known ancestors or descendants, but is interesting because it parallels many characteristics present in other apes and man.

also from middle Miocene deposits in Kenya, Turkey, Hungary, Russia, and China. Several species are represented over this wide area, but all are alike in having large flat-crowned, steep-sided molar teeth and rather small canines and incisors, features close to those of man.

It has been claimed that *Ramapithecus* had rounded tooth rows as in man and unlike the straight tooth rows in apes (see page 14), but recent work has shown this to be incorrect. In most respects *Ramapithecus* is ape-like with just a few hominid features superimposed.

Ramapithecus is still so similar to apes that it might seem strange to consider it a human ancestor at all. But since man evolved from the apes his very first ancestors would naturally be almost exactly like the apes of that time, the dryopithecines; and the problem is to identify the first adaptations which led the early men on to the path they are still following today. These first adaptations are thought to be those connected with small-object feeding. Alone among the Dryopithecinae *Ramapithecus* shows these features, and for this reason it is considered the most likely candidate for the ancestor of man.

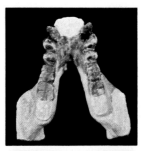

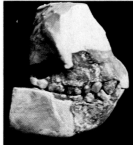

A reconstruction of the lower jaw (top) and lower face of Ramapithecus. Its ape-like features include its pointed canines, and front lower premolars; in contrast to these are the man-like large molars and small incisors and canines. These, with the heavy buttressing of bone evident on the upper and lower jaws, show that Ramapithecus used crushing and grinding movements in eating small objects.

Time Chart of Fossil Man

Names in bold type are fossil sites
Names in light type are cultures or culture sites
Sites in brackets are provisionally dated only

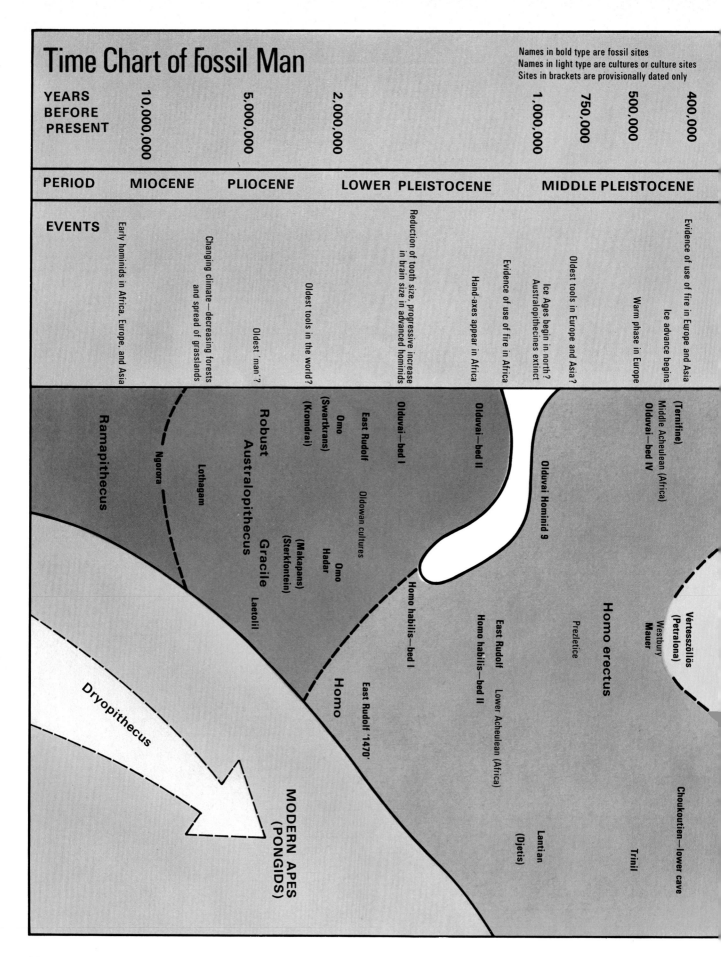

YEARS BEFORE PRESENT	10,000,000	5,000,000	2,000,000	1,000,000	750,000	500,000	400,000
PERIOD	MIOCENE	PLIOCENE	LOWER PLEISTOCENE			MIDDLE PLEISTOCENE	

EVENTS

Early hominids in Africa, Europe, and Asia

Changing climate—decreasing forests and spread of grasslands

Oldest 'man'?

Oldest tools in the world?

Reduction of tooth size, progressive increase in brain size in advanced hominids

Hand-axes appear in Africa

Evidence of use of fire in Africa

Ice Ages begin in north? Australopithecines extinct

Oldest tools in Europe and Asia?

Warm phase in Europe

Ice advance begins

Evidence of use of fire in Europe and Asia

Ramapithecus

Ngorora

Lothagam

Omo (Swartkrans)

(Kromdraai)

East Rudolf

Oldowan cultures

Oldvai—bed I

Omo

Hadar

(**Makapans**) (**Sterkfontein**)

Robust Australopithecus

Gracile

Laetolii

Oldvai—bed II

Homo habilis—bed I

East Rudolf

East Rudolf '1470'

Homo

Homo habilis—bed I

Homo habilis—bed II

East Rudolf

Oldvai Hominid 9

Prezletice

Lower Acheulean (Africa)

Lantian

(**Djetis**)

Homo erectus

(Terrafine)

Middle Acheulean (Africa)

Oldvai—bed IV

Mauer

Westbury

Vértesszöllös (**Petralona**)

Choukoutien—lower cave

Trinil

Dryopithecus

MODERN APES (PONGIDS)

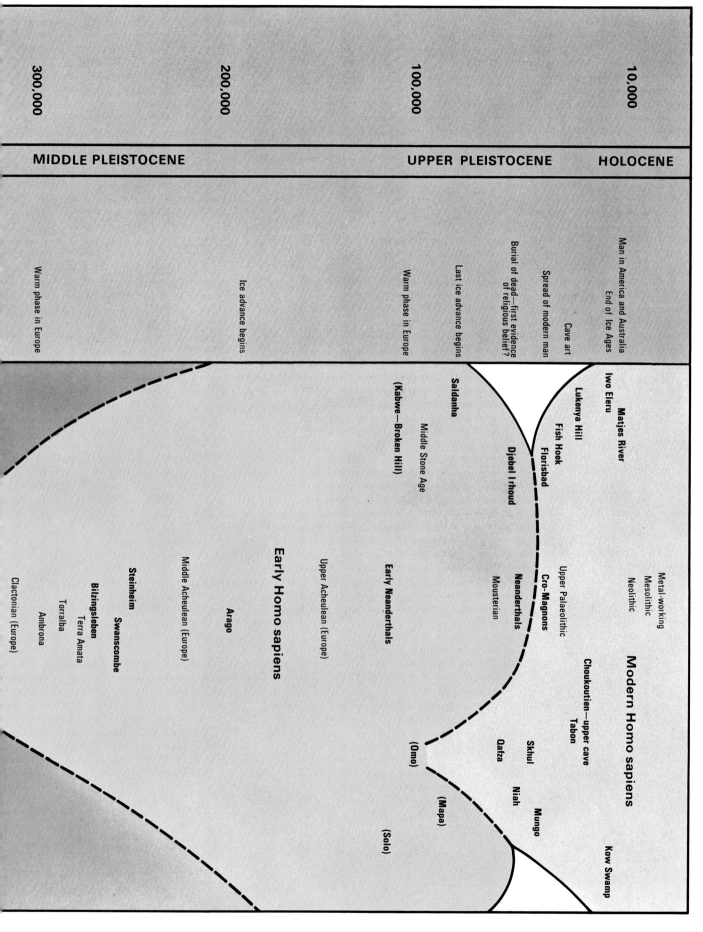

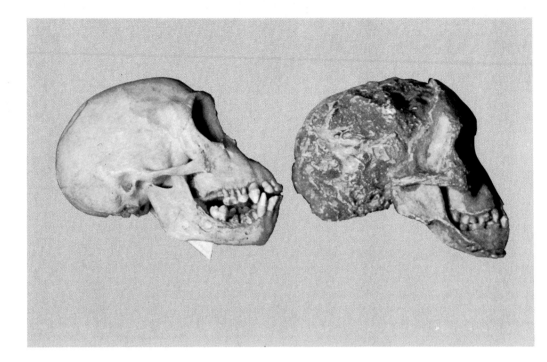

Left: A comparison of a cast of the Taung Australopithecus skull (right) with a young chimpanzee skull. The Taung child's skull has a natural cast of the inside of the brain cavity behind the face. Notice the dished face and small canines of the Taung fossil. Above: A cast of the Taung mandible (left) is compared with a mandible from a young chimpanzee. The Taung jaw has small canines and front teeth, and more human-looking first milk molars (behind the canines). The Australopithecine fossil shows large first permanent molars at the back of the jaw.

Before Adam

In 1925 Raymond Dart, a young professor of anatomy, described a fossil skull found a year earlier during mining for limestone at the Taung cave in South Africa. This was the first specimen of a group called the 'Australopithecines' (Southern Apes) which we now believe existed in various parts of Africa from at least five million years ago until about one million years ago and which contained forms directly ancestral to modern man, and other variants which became increasingly specialized and became extinct. The first find, classified as *Australopithecus africanus*, consisted of the skull of a child about six years old, which Dart regarded, from features of the skull, teeth, and face, as a real hominid – a member of man's own zoological family. But other authorities believed the find was a form of fossil ape not necessarily closely related to man. At that time fossils such as Piltdown man (see **page 34) were thought to be early human** ancestors, and the Piltdown specimen displayed a very advanced brain and skull with ape-like teeth and jaws, exactly the opposite combination to the South African fossils. Time has proved that Dart was

essentially correct in his conclusions about *Australopithecus*, and that those who believed in Piltdown man were cruelly hoaxed.

More discoveries were made by Dart, Robert Broom, and others in South African limestone caves at Sterkfontein, Makapans, Kromdraii, and Swartkrans. The first two caves produced finds which have now been classified as belonging to the same species as the original find, *Australopithecus africanus*, a lightly built early hominid, but the latter sites have produced fossils which appear to differ from this species in a number of respects, indicating a creature of larger build and with bigger jaws. Some authorities prefer to classify these other fossil specimens as a separate species called *Australopithecus robustus*, while others regard the difference as large enough for a separate genus called *Paranthropus*.

The various fossils of *Australopithecus* from South Africa have been recovered from caves which cannot be dated absolutely by radiometric methods such as potassium-argon. Comparisons of animal remains between the sites and with fossil

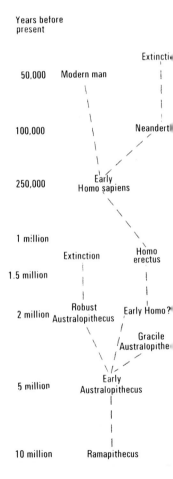

Years before present

Extinctio

50,000   Modern man

Neandert

100,000

Early
Homo sapiens

250,000

1 million

Extinction   Homo
erectus

1.5 million

Robust   Early Homo?
2 million   Australopithecus

Gracile
Australopithe

Early
5 million   Australopithecus

10 million   Ramapithecus

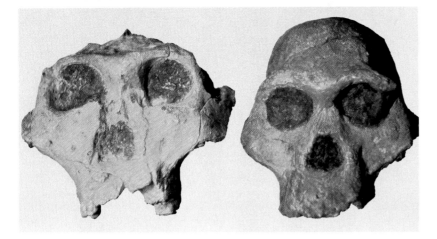

animals from elsewhere in Africa suggest that the most likely range of dates for these fossils is from over three million years old for sites such as Makapans to perhaps two million years for the later sites such as Kromdraii. Most animal remains from the South African sites come from open-country animals, which suggests an environment markedly different from that typical of apes. There is some evidence of stone tool-making from the sites but the tools cannot be certainly linked with *Australopithecus*. Dart put forward the idea that animal bones from the Australopithecine sites had been used as tools, but this has not been accepted by many anthropologists. It seems likely that many of the bone accumulations found in the caves were due to the activities of scavengers and predators such as the leopard. This might also explain why much of the South African fossil material consists of skull parts and neck vertebrae since these may have been severed from the carcass first, and then have fallen into the cave.

The lightly built *A. africanus* (or 'gracile *Australopithecus*') was probably only about 1·2 metres (4 feet) tall to judge by the known bones of the skeleton, and had a brain size of about 400 to 450 cc (compared with an average figure of about 400 cc for the chimpanzee and 1300 cc for modern man). *A. africanus* had teeth of essentially human type, without the large interlocking canines of apes, but larger than those of modern man and set in a robust jaw. The brain case was still small compared with the size of the face which was peculiarly flattened or 'dished'. The base of the skull was suggestive of a creature which walked upright, balancing its head

Above: A comparison of casts of skulls of a robust Australopithecine (Swartkrans SK 48; left) and gracile Australopithecine (Sterkfontein 5). Apart from the obvious differences in size and robusticity, the A. robustus *skull has a flatter top and a sagittal crest (bony ridge) down the middle.*

Right: Leopards often climb trees with their kills to escape the attention of hyaenas – and trees often grow around cave entrances. As the leopard feeds, parts of the kill may fall from the tree into the cave below. Much of the fossil South African cave material consists of skull parts, and neck vertebrae, which would have been the first parts to be severed from the carcass. Some of the bones at Swartkrans, including those of Australopithecines, show tooth marks which exactly match the canine teeth of contemporary leopards' jaws.

Casts of bones from East Rudolf and Olduvai which may come from the skeleton of Australopithecus boisei. From the left, a large robust humerus (upper arm bone), ulna (lower arm bone), femur (thigh bone) with typically small head and long flattened neck, and the upper part of a tibia (shin bone).

as man does (see page 14), since the area for neck muscle attachment was smaller than that of the large apes and faced downwards, and the foramen magnum, the hole through which the spinal cord joined the brain, was situated fairly centrally in the skull base, and again pointed downwards rather than backwards.

The whole skull of the stockier A. robustus was more massively built, with a **brain size of 500 to 550 cc. Several specimens have a sagittal crest (bony ridge)** along the top of the head rather like that found in a gorilla. This gives a larger area for the attachment of the temporal muscles to work the large lower jaws. But in the robust Australopithecines the crest is placed farther forward than in the gorilla, **and does not generally link with the bony ridge or nuchal crest at the back (marking the attachment of the neck muscles).** This perhaps indicates that the temporal muscles worked most strongly on the rear of the jaw in Australopithecus, where the molar teeth are situated. These teeth are certainly large in the robust Australopithecines and the front teeth seem relatively small – a feature even apparent in young individuals. The teeth are often heavily worn suggesting an abrasive diet, perhaps including seeds and roots, but the molars had thicker enamel to compensate for this. The lower jaws of the robust Australopithecines are generally more massive than those of the gracile Australopithecines, but also lack a bony chin.

The rest of the bones of Australopithecus

The Sterkfontein limestone cave site in the Transvaal is probably one of the oldest South African Australopithecine sites. Recent excavations there have produced many new specimens of Australopithecus africanus, and tools which may have been made by these creatures. The Swartkrans site is just over the horizon.

from South Africa are not known from complete skeletons, but looking at parts of individual bones we can piece together a picture of the body size and proportions of these creatures. Both forms of Australopithecus seem to have walked upright, to judge from the few hip and leg bones discovered, although the way in which they walked may have differed considerably from that of modern man. One pelvis from Sterkfontein has been reconstructed to suggest a creature of pygmy size, whereas the bones of the robust form suggest it was comparable in size to the modern human average. Estimates of the body weight vary but the gracile A. africanus may have weighed about 30 kilograms (65 pounds), and the robust form about twice as much.

Co-existence

We earlier suggested that certain South African cave sites contained either the gracile or the robust form of Australopithecus, and that the sites of the gracile form were earlier than the sites of the robust form. But some authorities believe that examination of larger samples of fossils shows that there was much greater variety in the Australopithecus populations from each cave. So perhaps the two forms co-existed or – it has even been suggested – the range of variation between gracile and robust is really due to a difference in sex – the gracile form being females and the robust form males of the same species! This is not now an acceptable explanation, at least as far as sites in East Africa are concerned. But what was the relationship between the two forms, if they really co-existed? If they lived in the same environment and led the same kind of life it is likely that competition between them would have soon resulted in extinction for one group, or its adaptation to a different environment. One suggestion, based on differences in the teeth, is that the robust form was a vegetarian, whereas the gracile form was a scavenger and small-scale hunter. The lack of really convincing evidence for tool-making and hunting behaviour by the Australopithecines from South Africa makes it impossible to answer this question with any degree of certainty, but looking at sites elsewhere in Africa there is much more evidence to help resolve the problem.

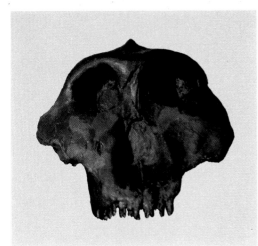

'Dear boy'

Olduvai Gorge in Tanzania is one of the many sites excavated by Mary and Louis Leakey (see page 49). Palaeolithic tools and many animal bones were known from Olduvai long before the first important hominid finds were made. There was great excitement in 1959 when a skull of a large Australopithecine, Olduvai hominid number 5 (O.H.5) was found on a 'living floor' associated with primitive pebble tools near the bottom of Bed I. Leakey named the find *Zinjanthropus boisei* (Zinj was an old name for East Africa, Boise was a man who had helped finance the excavations) but it is now accepted as another species of robust *Australopithecus* – even more massive than those from South Africa.

With the discovery of the 'nutcracker man' or 'dear boy' (as the Leakeys nicknamed O.H.5) it seemed that the maker of

Below: Louis and Mary Leakey at work (left). Leakey (see page 49) was one of the greatest figures in the study of early man. He was involved with discoveries of fossil apes or hominids from sites such as Rusinga Island, Fort Ternan, and, of course, Olduvai Gorge. Louis Leakey first visited Olduvai in 1931 and continued to be involved with work at the site until his death. His wife Mary often worked with him, and is still excavating in East Africa.

the early Oldowan pebble tools had been found at last. The sensation was all the greater when the relatively new potassium-argon technique was applied to volcanic rocks at the base of Bed I and showed that the new find was probably 1·8 million years old. This was far older than anyone had imagined possible for stone tools, the Australopithecines, or the early Pleistocene animal remains at Olduvai Gorge. These kind of dates are now almost commonplace for East African hominids, but at that time this single date resulted in a major revision of our ideas about African, and indeed world, prehistory.

Handy man

Soon after the *A. boisei* find there were more sensational discoveries at Olduvai when part of a lower jaw and some skull bones, a collar bone, and bones of a hand, a lower leg, and a foot were excavated from near the level of the previous find. These remains came from at least two individuals, one adult and one younger. The teeth in the lower jaw were clearly more advanced in the direction of modern man than those of *A. boisei* and from a reconstruction of the skull parts the brain size of the new hominid was estimated at about 650 cc. The skull parts were thin and showed no sign of the crests found in robust *Australopithecus* skulls. The foot bones indicated a creature quite well adapted to walking upright and perhaps about 1·2 metres (4 feet) tall, while the hand bones indicated the possession of a powerful grip. The Leakeys decided that

OLDUVAI GORGE

Olduvai Gorge cuts through the Serengeti plain in Northern Tanzania and exposes series of deposits laid down by a lake and by volcanic activity over a period of nearly two million years. There are six main formations in the Olduvai sequence of deposits from Bed I (the lowest and therefore oldest – early Pleistocene up to 1.8 million years old) to the late Pleistocene Naisiusiu Beds (formerly included in Bed V). Apart from the famous remains of fossil hominids and animals, Olduvai Gorge is important for the number of occupation sites excavated there, including living floors where bones or artefacts are recovered still lying undisturbed on the ancient land surface, and butchering or kill sites where artefacts are found together with the animal bones or even skeletons on which they were being used. Such sites have told us a great deal about the activities of early hominids beyond what we can learn from their own fossil bones.

this find represented the real ancient tool-maker at Olduvai; it was given the name *Homo habilis* ('handy man'), which showed the belief of Leakey and his colleagues that this creature was physically and culturally more advanced than *Australopithecus* and was actually human. (Other authorities have said that despite the larger brain size this is really an advanced kind of gracile *Australopithecus* and should be classified as either *A. africanus* or, if regarded as sufficiently distinct, *A. habilis*.) Further finds from deposits higher up at Olduvai have also been assigned to *Homo habilis* (including O.H.13, 14, and 24). These are clearly distinct from the robust *A. boisei*, which also seems to have continued to live at Olduvai until about one million years ago, and all the specimens of *Homo habilis* with cranial bones apparently exceed the cranial capacity of O.H.5. But some authorities feel that as the Bed II finds could be much later than the early Bed I finds, they could well represent a succeeding, more advanced species of true man, *Homo erectus*.

It has been suggested that the relationship between the two forms at Olduvai was one of hunter (*habilis*) and hunted (*A. boisei*), and this might indeed explain why the O.H.5 skull was surrounded by pebble tools. We cannot be certain at the moment but at least the Olduvai site establishes that the two forms of hominid did co-exist for a considerable period of time and must have had different ways of life. Whether both forms were capable of tool-making cannot be determined but there is good evidence of hunting activity from Olduvai and a construction of stones has been found – perhaps a hide for hunting, a windbreak, or the base of a hut.

More recently at a site called Laetolil, not far from Olduvai, Mary Leakey has found teeth and parts of jaws which are said to represent even earlier specimens of *Homo* dating from well over three million years ago. But these may well turn out to be examples of gracile Australopithecines when fully studied.

Omo and the Afar
The Omo valley in southern Ethiopia has been known as a site for early Pleistocene mammals for over forty years, but only recently has work there produced Australopithecine fossils and pebble tools

even older than those at Olduvai. Teeth, jaw, and skeletal fragments suggest that robust and gracile Australopithecines existed there over three million years ago. Northern Ethiopian sites of similar age have also produced early hominid fossils and tools; Hadar, in the Afar region, has yielded jaw fragments and nearly half of a complete and associated *Australopithecus* skeleton. This find, nicknamed 'Lucy', and further new discoveries of several more skeletons should tell us a tremendous amount about how the Australopithecines walked and lived. 'Lucy' is said to resemble the gracile Australopithecines, *Homo habilis*, and certain fossils from East Rudolf which we will look at shortly. The estimate of body size from the skeleton suggests a creature even smaller than previous estimates of the height of the gracile Australopithecines, although 'Lucy', as the name suggests, has been identified from the hip-bone as a female. Unfortunately there were only a few small skull fragments, but the teeth in the lower jaw seem 'human'. The bones of the skeleton may reveal the same story as those from other sites – that *Australopithecus* may have walked upright, but the body form and manner of walking was by no means identical to ours, although we do not have a satisfactory explanation for the differences at the moment.

The Lake Rudolf finds
Louis and Mary Leakey's son, Richard Leakey, has led several fossil-hunting expeditions to northern Kenya, close to where the Omo runs from Ethiopia into Lake Rudolf (now Turkana). Over two

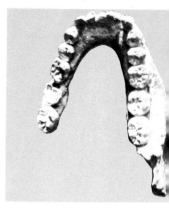

Above: The mandible of Olduvai Hominid 13 was found in 1963 during excavations in bed II. Associated with the lower jaw were skull fragments which suggest that the individual may have been a female with a brain volume of about 600cc. This mandible has been assigned to the species Homo habilis, although it is later and apparently more advanced than early Bed I finds such as the O.H.7 mandible (below). In some respects it appears comparable to advanced hominid specimens from East Rudolf such as number 992 (see page 59) and to later Homo erectus fossils.

Olduvai hominid 24 (nicknamed 'Twiggy' by the Leakeys) was discovered in 1968 in the lower part of Bed I, and although crushed when found, has been reconstructed into the most complete advanced hominid skull found so far from beds I–IV at Olduvai. Although of comparable age to the O.H.5 A. boisei skull (see page 57) this specimen is far more gracile, with a thin-walled rounded skull lacking crests, and with a small dentition. O.H.24 has been included in the species Homo habilis but shows resemblances to the gracile Australopithecines and certain recent finds from East Rudolf.

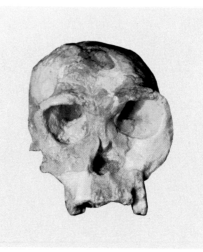

million years ago the lake extended farther east and rivers drained into it from that direction, laying down vast deposits now exposed in an arid landscape. A series of remarkable finds have been made in these deposits, and the first excavations by Leakey's team produced pebble tools and fossil hominid material referable to the robust Australopithecines, probably *A. boisei*.

For the first time there was enough material to identify clearly the sexes of the robust Australopithecines, and the male specimens are quite comparable to the *A. boisei* specimens from Omo and Olduvai. The female specimens are much smaller and have no sagittal crests, but can be matched with their male counterparts by the teeth (where present), or by the characteristic form of the bones round the ear region.

With the discovery of stone tools at East Rudolf, the question naturally arose of the identity of the tool-maker. Was there another *Homo habilis* living alongside the robust Australopithecines at East Rudolf? The question was answered in spectacular fashion by the discovery of find number 1470 in 1972. Hundreds of fragments were meticulously pieced together to produce a skull differing from anything else found before. The skull was relatively longer than other Australopithecine specimens and had a relatively large cranial capacity of at least 770 cc. This is within the range of later *Homo erectus* fossils and a large jump up from *Homo habilis* (cranial capacity about 650 cc), and an even bigger leap up from the Australopithecines, despite the fact that '1470' could be even older than any of the Olduvai fossils. Richard

A view of one site to the east of Lake Rudolf (now renamed Turkana) where some of the oldest known tools have been excavated by Richard Leakey and his team of workers. Below: A comparison of three humeri (upper arm bones). From the left: modern man; cast of East Rudolf find 739; gorilla. Note the robusticity and size of the Australopithecine fossil.

Right: Two mandibles showing the range of hominid types from East Rudolf. On the left is a cast of an advanced hominid mandible (find 992) from later deposits, similar to specimens of Homo habilis *and* Homo erectus. *On the right is a cast of a mandible of* A. boisei *(find 729) from earlier deposits but similar to specimens found in the same levels as 992.*

Right: The faces of East Rudolf finds 1470 (left) and 406 (right), both from deposits perhaps two million years old. The expanded brain case of 1470 contrasts with the flattened top and crest of the robust A. boisei *skull. The size of the tooth sockets indicated that the front teeth of the 1470 specimen were also relatively larger than those of the robust Australopithecines.*

Leakey and his colleagues have assigned this fossil to *Homo* but have left decisions about any species name until further studies have been completed. From East Rudolf deposits of the same early age other bones of the skeleton have been found which appear to fall into two groups. One group is like the known Australopithecine material, suggesting a creature with relatively long and strong arms and puny femur, while bones of the lower leg and foot suggest a biped of a different type from modern man. These bones have been assumed to belong to the robust Australopithecines at East Rudolf. But other bones, in particular several thigh bones, are much more like those of modern man and have been assumed to belong to a '1470' type of hominid.

The problems of classifying the hominids at East Rudolf have been made more difficult rather than more easy by the quantity of material found there. There are recent finds which in general cranial form resemble the gracile Australopithecines of South Africa and certain specimens of *Homo habilis* from Olduvai. The teeth, where present, seem remarkably human and like some specimens from Hadar. It will have to be decided whether these fossils represent another line of hominids, less advanced than '1470', or whether they could be within the range of a '1470' population with the latter specimen being a particularly large male, and the smaller specimens females. Yet another East Rudolf

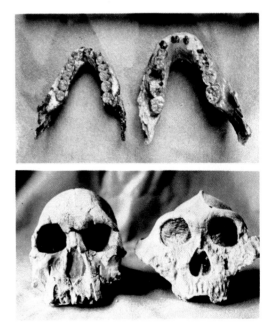

specimen has an advanced 'human' type dentition associated with a large upper jaw and face rather like the broad face of '1470'. Yet the brain case is smaller than that of '1470' and has a sagittal crest continuous with a nuchal crest! The presence of this crest has caused some confusion about how this find should be classified. But if the crest is regarded as a functional feature which appears wherever the jaws are relatively big compared with the cranium, then the crest becomes less important than other characteristics in classifying the hominid fossils. Accordingly this fossil may be judged from its teeth, face and cranial shape to be a variant of the 'advanced hominid' population.

The two distinct kinds of hominid bones found in earlier deposits at East Rudolf can be traced again in deposits dating from at least half a million years later. These later deposits, still over a million years old, have not produced much skull material, but there are excellent jaw fragments of

very robust Australopithecines, some even more massive than those from Omo, Olduvai, or the earlier East Rudolf finds. But there are also hand-axe tools, and bones including a well-preserved lower jaw of a very different creature, indistinguishable in the parts preserved from the Pleistocene species of true man, *Homo erectus* (see page 63). Recently it has been announced that a skull of *Homo erectus* type has also been found. This species had spread over much of the Old World by about a million years ago, and by then was probably the sole surviving hominid species, since the remaining Australopithecines had become extinct.

How is it possible to make sense of the evolution of the Australopithecines, with at least two different forms co-existing in South Africa and East Africa? Australopithecines combine typically hominid teeth (clearly highly evolved from a Miocene ancestor such as *Ramapithecus*), a skeleton adapted for a kind of bipedalism,

In this reconstruction a small group of Australopithecus robustus *(left) peers over a rocky outcrop at a settlement of '1470' hominids. Fossil remains of both these have been found in deposits of the same age at East Rudolf. The evidence suggests that the more advanced and lightly built '1470' hominids were tool-makers and meat-eaters, while A. robustus was a less aggressive vegetarian. Hominids of these two types seem to have existed together for about half a million years. '1470' has been placed in the genus* Homo *but so far no decision has been taken on a species name.*

and a brain of ape size although not necessarily of ape-like structure. One suggestion is that scavenging and hunting behaviour was an early feature of the hominid line where tool use, subsequent reduction of the canine teeth, and the development of bipedalism set in motion a rapid divergence from the evolving Dryopithecine apes. But if this were so one might expect bipedalism and tool-making to have advanced beyond the stage represented by the Australopithecines, given a development period of perhaps ten million years. The fact that *Ramapithecus* in the Miocene already had significantly reduced front teeth and broader back teeth surely cannot have been due to developed tool-making at that early date. It is now thought more likely that the initial hominid divergence occurred through dietary adaptations to the increasing open-country environments developing in Africa and elsewhere during the later Miocene. Unfortunately there is only fragmentary fossil material from African sites such as Ngorora (perhaps nine million years old) and Lothagam (probably over five million years old) to help us understand what was occurring during the time gap between the possible ancestral form *Ramapithecus*, and the later Pliocene *Australopithecus* material.

Because of the lack of fossil material to fill these crucial gaps in our knowledge of hominid evolution scientists have turned to modern day primates and man to help find clues. Dr Clifford Jolly studied gelada monkeys which live in arid areas of Ethiopia and found that the geladas differed from typical baboons in ways which corresponded to the differences between hominids and apes. He suggested that this situation had arisen because hominids, in their divergence from apes, had undergone some parallel adaptations to those made by geladas in their divergence from the baboons. These adaptations were concerned with living in a rather inhospitable environment where food resources were scattered, and where the diet consisted mainly of small hard food objects such as seeds of grasses. The 'seed-eating hypothesis' (see below) would place *Ramapithecus* as an early member of the hominids, already showing the hallmarks of its adaptations by its smaller front teeth and broader back teeth. The robust Australopithecines would represent fully developed 'seed-eaters' with the later

specimens in East Africa (*A. boisei*) being the most extreme in their specialization prior to their extinction about one million years ago. The gracile Australopithecines would perhaps represent a second phase of hominid adaptation in which certain populations were switching to a diet with more meat, and where hunting and carrying activities would impose a progressively greater 'human' set of adaptations on the earlier seed-eating ones. *Homo habilis* and '1470' might be even further developed members of this evolving line.

But a big problem with the 'seed-eating' theory is the lack of a plausible explanation for the origin of bipedalism. Looking for predators over tall grass still does not seem a good enough explanation, although once bipedalism appeared it would obviously be useful for carrying food and weapons.

Below left: Dr Clifford Jolly's 'seed-eating' hypothesis shows how Miocene apes could have evolved into hominids as a result of adaptations to a new diet. The blue boxes in this diagram show the supposed characteristics of the Dryopithecines – leaf-eaters with gorilla-like teeth and the ability to swing the arms. Jolly assumed their behaviour to be like the chimpanzee's, with large social groups and occasional use of natural 'tools'. Their way of life had to change as the forests gave way to arid, grassy plains, where the main form of food available was grass-seeds. The buff boxes show adaptations that may have occurred. Front teeth, no longer needed for tearing, become smaller in relation to the powerful back teeth, and the jaw shape alters in proportion. A large, mobile tongue moves the seeds to be crushed by the molars. The creature now sits upright to feed, and uses its hands – a first step towards bipedalism. Small groups form within the main group, in response to the need to go farther afield for food. Each group is dominated by a single male. The row of grey boxes show Australopithecine characteristics, the result of these adaptations. The green boxes show the next modification – adaptation to the life of a hunter. A diet based on meat changes the teeth again, as the crushing function of the molars becomes less important. Hunting for the meat demands that the creature's rather shambling walk becomes more efficient. Behaviour also evolves; the use of found objects gives way to the manufacture of weapons and cutting and carrying utensils; communication becomes important for members of hunting parties, and large projects are shared with other troops. The yellow boxes represent the characteristics of modern man.

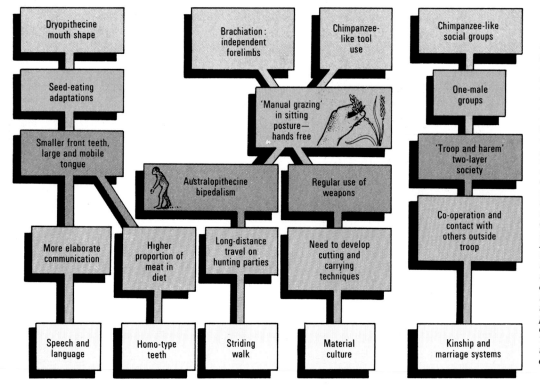

62

The First True Men

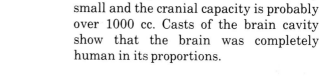

The genus *Homo*, man, had appeared in the fossil record without any doubt by one million years ago. The Australopithecines had become extinct or had evolved into a form of man which we call *Homo erectus*, 'erect man'.

The first fossil remains of *Homo erectus* were found in 1891 at Trinil in Java, in river deposits which are now regarded as over half a million years old. They included the top part of a skull, and some femora (thigh-bones) which were very similar to those of modern man. In fact apart from a bony growth on one femur due to injury, the leg bones look so modern that it has been suggested that they do not belong with the 'primitive' skull-cap but had somehow become mixed with it in the river deposits. The skull-cap is thick-boned and very flat on top with no sign of a forehead development. It has a thick brow ridge (supraorbital torus) above the eye sockets and a brain capacity of only about 850 cc compared with a modern average of about 1300 cc.

Further finds from Java give us a picture of *Homo erectus* over perhaps a million years of human evolution. Finds from the lower Djetis beds are probably over one million years old, and the Modjokerto site from that time period has produced the skull of a child as well as a massive upper jaw and back part of a skull. These remains are undoubtedly of *Homo erectus*, but there are also a few intriguing remains of teeth and large jaw fragments which have been given the name *Meganthropus* ('giant man'), and these might represent an Australopithecine form co-existing with a more advanced hominid – an interesting parallel with evidence from Africa. The fossils from the lower and middle Pleistocene of Java show an apparent evolutionary trend from the earliest to the latest specimens, with a reduction of the size of the teeth and jaws, and an increasingly large brain housed in a higher and less rugged skull. One of the most recent finds is a whole skull, partly distorted but displaying, for the first time

This composite reconstruction of an early Homo erectus *skull from Modjokerto, Java, was made from parts of upper and lower jaws, and the back part of a skull. The rugged face is more massively built than the later Pekin specimens but compares well with more complete finds of* Homo erectus *material from East Rudolf (Turkana) and Java.*

Some of the principal sites where fossil hominids have been found.

from Java, a fairly complete and massive *Homo erectus* face. But it is quite advanced in other respects – the teeth are fairly small and the cranial capacity is probably over 1000 cc. Casts of the brain cavity show that the brain was completely human in its proportions.

Pekin man

From Lantian in China come a few finds which may be of a similar age and form to the more rugged Djetis remains from Java, but the most famous Chinese remains of early man are those from the lower cave at Choukoutien, near Peking. The existence of 'Pekin man' was first suspected from fossil human teeth found in

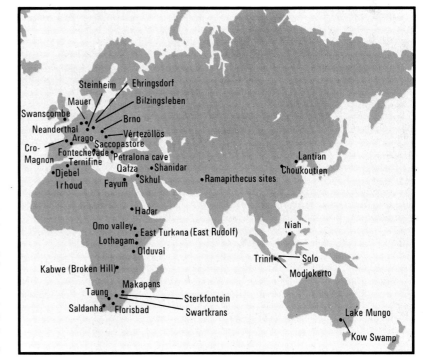

apothecaries' shops. The Chinese used to grind up fossils to use as medicines since they believed the fossils were dragons' bones and teeth with magical properties. Some of these teeth were traced back to the Choukoutien site where excavations were carried out producing quartz tools and then more human teeth. Professor Davison Black gave the name *Sinanthropus pekinensis* ('the Chinese man of Peking') to one tooth, and the name was subsequently given to all the human fossils from the lower cave at Choukoutien. There were remains from about 40 individuals but the whole collection, except for the one tooth, was lost in China during the Second World War. Even now there are rumours that the collection still exists in Japan, China, or the United States, but it is most likely that all that now remains to science are the studies on the original material and fine sets of casts of the bones. From these it is clear that the Pekin remains are very similar to those from Java and should also be reclassified as a type of *Homo erectus*.

Some of the broken human bones had been burned in hearths just like the many broken deer bones found there, so it seems possible that the Pekin people were cannibals; but the main foods of Pekin man seem to have been animals such as deer, elephant, and rhinoceros, and fruits such

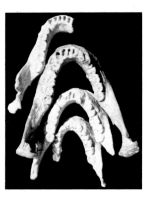

This comparative picture shows three 'Atlanthrops mauritanicus' mandibles from Ternifine in Algeria and, below, a cast of the O.H. 13 Homo habilis mandible from Olduvai. The three North African Homo erectus mandibles display a wide variation in size but all have large teeth and are reminiscent of specimens from Choukoutien in China. Right: This composite reconstruction of a female skull was made from several specimens but probably gives a quite accurate impression of the less rugged Pekin material. Although the brows are still thick and wide there is more development of the forehead region than in the Java specimens, and a general expansion of the braincase associated with a bigger cranial capacity.

as the hackberry. Excavations at Choukoutien have been restarted recently and have already produced more skull parts, teeth, and limb bones, so we can probably look forward to more evidence from China about *Homo erectus*.

Africa

The appearance of hand-axe (Acheulean) tools (see page 92) made from lava in Upper Bed II at Olduvai Gorge led the Leakeys to search for the remains of the people that made them, and from discoveries made elsewhere in the world it seemed likely that *Homo erectus* remains would be found there too. This possibility was confirmed in 1960 by the discovery of a skull, perhaps one million years old. Although this skull, Olduvai hominid 9, has not yet been fully described it has a general resemblance to *Homo erectus* skulls from Peking and Java. It is flat-topped, thick-walled, and broad across the base with massive brow ridges, but has a brain capacity of over 1000 cc. Part of a very robust pelvis and a femur have also been found higher up, in Bed IV, and might be about half a million years old.

This femur has some interesting resemblances to those of Pekin man who was living at about the same time 7000 miles away. So, did the African and Asian *Homo erectus* people come from the same stock, and if so, where did they originate? One theory is that these peoples evolved outside Africa and came into that continent with hand-axe cultures, which seem to have arrived rather abruptly in the Olduvai sequence. The hand-axe cultures

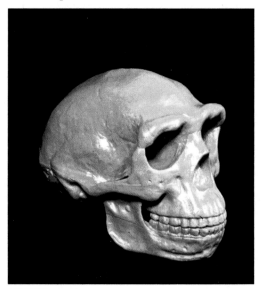

must then have co-existed for a time with developed pebble-tool cultures (perhaps made by evolved Australopithecines).

It seems more likely that *Homo erectus* developed in Africa, since the earliest hand-axe industries are known from African sites including deposits at Ileret, East Rudolf, and are over one million years old. As we have already mentioned it seems that a form of true man, perhaps an early form of *Homo erectus*, co-existed with the last of the robust Australopithecines at East Rudolf. Some rather fragmentary fossils from East Rudolf and Olduvai suggested that a hominid with a thicker-walled skull, larger brain capacity, and larger brow ridges could have evolved from forms such as '1470' and *Homo habilis* during a period of nearly a million years. Recent, more complete discoveries have confirmed this. We are still unable to explain the changes in skull form which must have occurred between two million and one million years ago to produce the *Homo erectus* cranial structure. Perhaps the change is related to an increase in body size and ruggedness that came with a transition from the small-scale hunting and scavenging which produced meat for the Australopithecines, to large-scale hunting of large animals (of which there is much evidence from Olduvai and East Rudolf).

Europe

Evidence of *Homo erectus* from Europe is less complete but more controversial. Found in 1907, the lower jaw from the Mauer sand pit near Heidelberg in Germany combines a very thick, chinless body with teeth which seem almost too modern to be half a million years old. Nothing that can definitely be called a tool has been found at Mauer although primitive tools of similar age are known from other European sites such as Westbury-sub-Mendip in England. Primitive tools of an even earlier age are known from Prezletice near Prague in Czechoslovakia but unfortunately the only probable human fossil is a tiny fragment of a molar tooth. Tools reminiscent of African pebble tools have been found at the site of Vértesszöllös in Hungary, dating from about 400,000 years ago. Hopes of finding remains of the people who camped and lit fires at the site were realized when an occipital bone and some

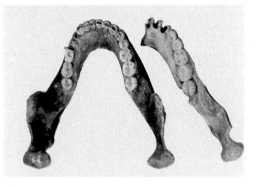

Right: Casts of the Mauer (Heidelberg) mandible (left) and the larger of two mandibles from Arago (Tautavel). The Mauer jaw, over 250,000 years older, appears relatively advanced in its dentition, but part of the size difference may be due to the fact that the Mauer fossil could be from a female, and the Arago mandible that of a male.

At the Spanish site of Ambrona (200 km north of Madrid), excavations have revealed evidence of the activities of early Acheulean hunters of perhaps 300,000 years ago. The tools used for butchering large mammals were found scattered over the site and it seems that fire may have been used to drive herds of animals into swampy country where they could be attacked more easily. An unusual arrangement of elephant leg bones could have been made by the hunters for stepping stones.

teeth were found there in 1965. The teeth are apparently like those of *Homo erectus* from Choukoutien – but the occipital bone, although thick, is larger and somewhat more rounded than other *Homo erectus* specimens. The brain capacity of this hominid must have been well within the modern range (over 1300 cc). So some authorities prefer to call this a *Homo sapiens* occipital, implying that it belongs to the same species as modern man; but as we shall see the problem of drawing up a dividing line between *Homo erectus* and *Homo sapiens* is not easy. If, as we believe, the former species evolved into the latter, there must have been a long period of time when people with intermediate characteristics existed.

All of these sites have provided some information about the environment in

which early man lived, and some show that as man moved from the warmer tropical areas he took with him the use of fire (and perhaps crude clothing). Other sites such as Torralba and Ambrona in Spain, and Terra Amata in France (perhaps 300,000 years old), have yielded much evidence of the way of life of European early man, but no hominid remains (except for one foot-print). Terra Amata was a campsite; it has revealed that the late *Homo erectus* or early *Homo sapiens* people of this time constructed quite sophisticated dwellings apparently using wooden poles and stones to support a roof made of hide or branches. The dwellings formed seasonal camps probably inhabited by 15 or 20 people, and each contained a hearth and 'feeding area'. Fire may have been used to drive animals into areas such as swamps where they could be hunted more easily. The ability of these hunters is demonstrated by the number of straight-tusked

elephant, horse, wild boar, oxen, and rhinoceros bones butchered and scattered with stone tools over these sites.

Some of the peoples of this period were makers of early hand-axe tools, others made tools such as the kind found at Clacton, which do not include hand axes (see page 92). Remains of people who made Clactonian type artefacts have recently been discovered at Bilzingsleben, East Germany; they include an occipital bone quite similar to those from Vértesszöllös, Petralona, and Peking, suggesting that some peoples in Europe were still similar in many respects to *Homo erectus*. Slightly later (perhaps 250,000 years ago) people making quite sophisticated hand-axe tools lived near the banks of the river Thames in England (then running on a different course across the land-link with Europe as a tributary of the Rhine). At one site, Swanscombe, two parietal bones and an occipital bone have been found which fit together to produce the back part of a skull which is quite advanced compared with

Neanderthal man competed for shelter with cave bears, the remains of which have often been found in caves showing traces of his occupation. It seems possible that the Neanderthals had a cave bear cult, since the bears' skulls have been found formally arranged.

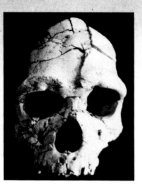

A facial view of the 'Tautavel skull' from the Arago site in the French Pyrénées. Note the large brows and broad face and nasal opening. This fossil is probably that of a young adult male.

Left: A European specimen which seems to show characteristics of both Homo erectus *and* Homo sapiens *is the skull from the Petralona cave in Greece, probably about the same age as the Mauer and Vértesszöllös fossils. Although the brain size was relatively large, the massive broad face, large brows, angulated occipital bone and broad base to the skull are all reminiscent of* Homo erectus. *Although the teeth present are relatively small the lower jaw must have been even bigger than the Mauer jaw.*

that of *Homo erectus*. Many people have regarded it as one of the first remains of *Homo sapiens*. But its modernity has probably been over-emphasized since it is still fairly thick and low in height, with shorter, flatter, parietal bones and a wider occipital bone than modern *Homo sapiens*. A skull of the same date or slightly later, the Steinheim skull, was found in river deposits near Stuttgart in Germany. In its general form it seems to represent a significant advance over earlier *Homo erectus* specimens but still has large brows and a small cranial capacity (less than 1200 cc.). Part of its difference from earlier fossils may be due to difference in sex.

This view is supported by the finds from the Arago (Tautavel) cave in the French Pyrénées. Recent excavations there have produced parts of lower jaws, teeth, and most important, the face of a young adult (probably male) dating from the beginning of the Riss glaciation, perhaps 200,000 years ago. The 'Tautavel skull' could well be a male equivalent of Steinheim and although the breadth of the upper face is large, as are the brow ridges, the profile of the face appears fairly modern and the teeth in the upper jaw are not large. The two jawbones from Tautavel reveal a difference in size probably due to difference in sex.

Fossils from the last interglacial (120,000 to 70,000 years ago), such as

those from Ehringsdorf in Germany and Saccopastore in Italy, can be regarded as ancestral to the Neanderthal people (see below). There are some fragmentary fossils of the same age from Fontéchevade in France, which some authorities believe to be of a different type, an ancestor of modern man.

Was there then more than one group of early men living at the same time in Europe during the Pleistocene? The fact that some of the European fossils are associated with hand-axe industries and others with industries which usually contain no hand-axes could indeed mean that there were varied races of men in Europe, some more modern than others, but it is also possible that differences in the tool types found at various sites may be partly due to differences in the activities carried out at the sites or to a difference of environment.

We know from sites such as Tautavel that there was much variation within human populations at this time, and statistical analyses of the fossils suggest that all these specimens could perhaps belong to one evolving line.

The Neanderthals

During the coldest stages of the ice ages *Homo erectus* and early *Homo sapiens* populations do not seem to have managed to survive in Northern Europe. Probably

they had not yet developed a culture that could adapt to cold, even though they had the use of fire. But in the last Ice Age, which began about 70,000 years ago, some human groups did manage to stay behind to live off the plentiful herds of large mammals such as the woolly rhinoceros and mammoth, rather than be pushed southwards before the advancing tundra environment. These resourceful people were the Neanderthals (*Homo sapiens neanderthalensis*), who managed to survive in severe conditions from 70,000 to 35,000 years ago by making the best use of natural shelters such as caves, and by improving cultural adaptations and hunting techniques to cope with the new environment. The short, stocky build of the European Neanderthals shows that they must also have become adapted physically to the periglacial environment through natural selection.

The first remains which were recognized as belonging to this palaeolithic race were found in the Neander valley near Dusseldorf in Germany in 1856. Many more finds have been made since, from Spain, Italy, and Israel to the British Isles, Czechoslovakia, and the Soviet Union. The skulls are long and low with large brow ridges

From the anthropologist's point of view one of the most fortunate cultural developments of the Upper Pleistocene was the custom of burying the dead. This has led to whole skeletons of Neanderthal and early modern people becoming fossilized in caves. This reconstruction combines evidence of the burial practices of the Neanderthals from several different sites. Graves were generally small and shallow; the body often lay with bent knees. Wild flowers are thought to have been strewn on the body of the Neanderthal man buried at Shanidar, Iraq, while animal bones or flint tools have been found at sites such as Teshik-Tash, USSR, and La Chapelle and La Ferrassie, France. The burials at La Ferrassie include a man, a woman, and four children, arranged in a complex pattern.

Below: Viewed from the side the Neanderthal face projected around the nose and was rather flat below. The brain of the Neanderthals was particularly wide in the lower parietal region, giving the skull an almost spherical shape when viewed from behind. The occipital region bulged out, although it was not angulated as in Homo erectus *and lacked the strong torus found in earlier forms. This skull was found in a quarry at Gibraltar in 1848 but its importance was not recognized until long after the Neanderthal discovery in Germany in 1856.*

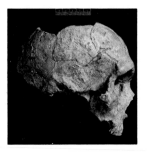

which curve over the orbits (eye sockets) and are less broad than in *Homo erectus* and less thick at the sides of the face. The average size of the Neanderthal brain was far greater than that of *Homo erectus*, being even larger than the modern average and exceeding 1600 cc in several specimens. The frontal lobes of the brain were not as large as in modern man, but were expanded enough to need a broad frontal bone so that Neanderthals do not have the 'pinched-in' appearance of *Homo erectus* skulls.

The Neanderthal face was particularly distinctive. This unusual form was probably an adaptation to the climate. It was dominated by a projecting and voluminous nose with swept-back cheek-bones – probably to cope with breathing extremely cold air and to protect against dangers such as frostbite. The eye sockets were high and more circular than the rather rectangular orbits found in many earlier fossil men. There was no canine fossa – a hollowing of the cheek-bone found in modern man and some earlier forms of hominids – and the sinus (or air-space) in the cheek-bones was well developed.

Neanderthal teeth were smaller than those of earlier fossils but larger than the modern average, particularly the front teeth. Neanderthal lower jaws sometimes showed a development of a bony chin which is a modern characteristic. The Neanderthal skeleton, at least in the European specimens, suggests a short and very muscular build, with some individuals only about 1·5 metres (5 feet) tall. There is no evidence that Neanderthal man could not walk upright, and early studies suggesting that the Neanderthals had grasping toes like apes and a stooping walk were mistaken and based in one case on a Neanderthal skeleton suffering from arthritis!

There was certainly nothing backward about the way of life or cultural ability of the Neanderthals. 'Mousterian' industries (named after one of the best-known Neanderthal sites in France) were of many different sorts. There is plenty of evidence of the Neanderthals' concern for their dead from ritual burials containing grave goods, which seem to indicate a belief in an after-life. Neanderthal people in the Near East apparently even laid flowers in graves of their dead, as revealed by pollen analyses from the Shanidar cave in Iraq. The Neanderthals from the Near East seem to have been rather less specialized than their European relatives. They were closer to modern man in certain respects, being taller and less robust in their skeletons, and with certain more modern features in the skull. But in general skull shape, brain size, and form of the shoulder-blade and pelvis they were unmistakeably Neanderthals. Their culture was a form of Mousterian called the Levalloiso-Mousterian. Other remains showing Neanderthal and modern characteristics are known from Djebel Irhoud in Morocco.

Early modern man

From the Skhul and Jebel Qafza caves in Israel there are burials associated with Levalloiso-Mousterian artefacts, which appear to be of about the same age as the Neanderthals (or in the case of Qafza, perhaps even earlier) and yet much more modern in their anatomy. The best preserved specimens show a high, relatively short skull, and the parts of mandibles preserved show well-developed bony chins. In fact the skulls and skeletons seem to be just very robust examples of modern man, *Homo sapiens sapiens*.

What was the relationship between Neanderthals and these modern men? One theory suggests that the first modern men actually evolved from the Neanderthals in the Near East, or in many places including Europe, and that improvements in cultural ability led to a reduction in the size of the Neanderthal teeth and face, giving rise to a modern skull form. But if these modern men were in fact associated with Mousterian-type tools and had teeth as big as or even bigger than the Neanderthal average, it seems unlikely that cultural improvement led to their evolution from the Neanderthals. And these Near Eastern robust modern skulls seem rather more like the earlier European fossils than the later 'specialized' form of Neanderthal man. Are there any other possible ancestors for the Near Eastern modern men if they did not evolve from the Neanderthals?

Africa and Asia

There are fossils from Broken Hill in Zambia, Eyasi in Kenya, and Saldanha

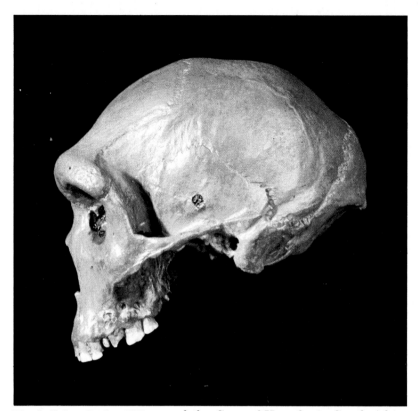

The skull from Broken Hill (now called Kabwe) in Zambia (formerly Northern Rhodesia) was found during mining operations in 1921. It is characterized by a very broad and thick brow-ridge, a flat frontal bone, an angulated occipital bone with a strong torus (mound of bone) running across it, and a large face. But other features such as the form of the parietal bones and base of the skull are very like modern Homo sapiens.

and the Cave of Hearths in South Africa which are probably 70,000 to 120,000 years old and seem to represent a distinct population of early *Homo sapiens*. The Broken Hill skull is particularly well preserved and displays a mixture of *Homo erectus* and *Homo sapiens* characteristics, reminiscent of European fossils such as Petralona and Tautavel. It would be a mistake to regard these fossils from south of the Sahara as merely African Neanderthals. They seem to represent a quite different form of early *Homo sapiens*. But if these fossils are only about 100,000 years old they seem too late in time to be ancestral to modern man, although some authorities believe this was possible. More likely ancestors for modern man are some

Side view of the Qafza (also spelt Kafzeh) skull number 6, found in 1934 in Israel. Recent excavations have recovered several new specimens. The skulls are characterized by a high rounded brain case, and although the teeth are large, the face is flat in its middle part with a distinct canine fossa (hollowing in the cheek bones), only projecting below the nose.

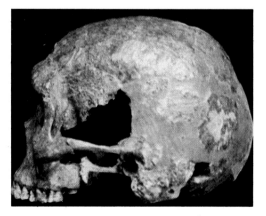

This skull (Omo 2), discovered in the Omo valley, Ethiopia, in 1967, displays a remarkable combination of advanced and primitive features. The brow-ridges are only slightly developed, yet the occipital bone is strongly angled with a robust occipital torus like that found in Homo erectus skulls. The other main find from this area – Omo 1, a skull and skeleton – is even more modern in its features, although both finds may be over 100,000 years old.

The Saldanha remains, the top part of a skull and a fragment of lower jaw, were found near Hopefield in South Africa in 1953. They are similar in many respects to the Broken Hill specimen, and could represent a female individual of Homo sapiens rhodesiensis.

A side view of one of the skulls from Solo, Java, which shows a close resemblance to the earlier Homo erectus skulls. It has large brows, flattened frontal bone and skull, and an angulated occipital bone with a strong torus (ridge).

fossils from late Middle Pleistocene or early Upper Pleistocene deposits in the Omo Valley of Ethiopia. The two best specimens are partial skulls, one of which has an associated skeleton. One shows a remarkable mixture of *Homo erectus* and *Homo sapiens* characteristics, having a slight brow-ridge development but a strongly angled occipital region with an occipital torus, very like certain *Homo erectus* specimens. The height of the skull and the cranial capacity seem quite modern in both Omo skulls, as is the form of the temporal bone and mastoid process. The second skull and associated skeleton are very modern looking, but large and robust, resembling the Skhul and Qafza fossils in some respects. The broad form of frontal bone is similar to the other Omo skull, but the back of the skull is really very modern. The associated lower jaw possesses a bony chin. At the moment it is not possible to say whether these two skulls belong to one very variable population of fossil men, since they were found about a mile apart. But the more modern-looking skull is certainly a good candidate for an ancestor of modern man, especially if a suggested age of 120,000 years can be confirmed. The ancestors of the Omo hominids are unknown, although there are resemblances to some earlier hominids such as Vértesszöllös, Swanscombe, and Salé in Morocco – a new skull showing characteristics of both *Homo erectus* and *Homo sapiens* was found there in 1971.

From the Far East come some fragmentary Chinese fossils which could be descendants of the earlier Pekin *Homo erectus* man. Perhaps they are eastern equivalents of the Neanderthals, or they may be related to the Upper Pleistocene remains from the Solo River in Java. Eleven partial skulls were found there in deposits which are uncertainly dated but which could be about 100,000 years old. The skulls were found without remains of the face, teeth, or any bones of the skeleton except for two shin-bones, and it has been suggested that these fossils represent the remains of a cannibalistic meal where the skulls had been broken open to get at the brains within. The Solo fossils are very reminiscent of the more ancient Java *Homo erectus* fossils, and it is likely that the Upper Pleistocene Solo people were direct descendants of these Middle Pleistocene *H. erectus* forerunners. The main difference is that the Solo brain-case is bigger – with an average capacity of over 1100 cc.

Some authorities have suggested that this South-east Asian line continued to evolve and gave rise to the modern Australian aborigines. Similarly they would derive modern Negroes from the Broken Hill people in Africa, and modern Europeans from the later Neanderthals. This seems unlikely since the appearance of modern man in many areas was apparently much earlier than previously supposed (see page 73) and races of modern man are very closely related to each other. This suggests that the modern races did not have a long separate ancestry in different parts of the world. We can expect that further discoveries will be made in the next few years which will sort out some of the problems of the ancestry of modern man, but it seems that the Neanderthals in Europe and the Near East became more distinct physically from modern man as time went on, rather than approaching more closely to the modern type. Between 45,000 and 30,000 years ago the glacial climate of Europe underwent at least two warmer fluctuations, which may have led to the influx of a modern type of man from South-west Asia who was culturally more advanced than Neanderthals and who may have had an advantage over them in their period of co-existence which lasted for several thousands of years.

The Cro-Magnon people lived in rock shelters, often with inner tents of skins. Their clothes were of scraped skins, roughly sewn, and they used tools of stone and bone. They also made bone ornaments. Their magnificent paintings were executed not in the areas where they lived, but deep inside cave systems.

The Cro-Magnons

These people were the 'Cro-Magnons' (named after a French site which produced some early important discoveries of them) and they occupied cave sites and open sites in Europe from over 30,000 years ago to 12,000 years ago. The upper palaeolithic cultures associated with the Cro-Magnons were quite distinct in technique from those of the Neanderthals, and much greater use was made of bone, ivory, and antler. One, the Aurignacian culture, spread from eastern to western Europe; it may perhaps have developed in the Near East. The other early upper palaeolithic industry, the Châtelperronian (or lower Périgordian) has some links with one particular kind of Mousterian and this may mean that there was some contact between the last Neanderthals and the first modern men in Europe. As well as the Châtelperronian and Aurignacian there were later distinctive European cultures such as the upper Périgordian, Solutrean, and Magdalenian. The remains of the people who made the various industries indicate a corresponding physical diversity in the Cro-Magnons, some of them being short and stocky and others much taller and long-limbed. If they derived from the earlier robust modern peoples of the Near East or North Africa such as Qafza or Omo, it might be asked why it took so long for modern man to occupy Europe and the rest of the world. One answer, in Europe at least, is that the Neanderthals must have been exceptionally well adapted physically and culturally to survive there at all, and perhaps only the temporary climatic improvement and subsequent migration of animals led modern peoples to move into Europe between 45,000 and 30,000 years ago. We must also remember that for all their modern appearance, the Qafza, Skhul, and Omo men did not have tools of upper palaeolithic type, and perhaps it was only with the development of the most sophisticated palaeolithic tools that people became able to shape different environments to suit their needs, finally giving the impetus for the spread of modern man.

71

Man's Migrations

All modern men are unquestionably members of one species, *Homo sapiens*. But within our species it is possible to recognize populations which differ from one another in the rate of occurrence of certain inherited characteristics controlled by *genes*. Such populations are known as races. Many different classifications of races have been proposed in the past; these classifications are all simplifications of a complex situation since there will always be some groups which do not fit exactly into the categories used, and apparently distinct races often merge gradually where their ranges meet.

Characteristics or traits which can be measured and used in classifying races include skin colour; hair colour, form, and distribution over the body; the structure of the skin-folds above the eyes; the shape of the nose and lips; general physique; and types of blood groups and fingerprint patterns. Many of these characteristics appear to be affected by the action of natural selection in differing environments, so it is not surprising that, as human groups spread out into the remarkable range of habitats in which they are found today, they became physically different from one another.

One convenient division of modern man distinguishes six races: *Caucasoid* or 'white' (actually quite varied in skin colour); *Australoid*; *American Indian Mongoloid*; *Asian Mongoloid* (also including the Eskimos of North America); *Negroid*; and the *Bushman* or *Capoid*. There are as well several problematical populations such as the Ainu of Japan, the Polynesians, and the Andaman Islanders of the Indian Ocean, who do not fit easily into any of these categories. Each of the six racial groups has some distinctive bodily features which may be recognized in the skeleton, and these can be useful in finding out when such racial differences first became established. Where there is little fossil evidence it is possible to use similarities in genetic factors (such as blood groups) or cultural factors (such as languages) to link populations who were once closely related but have migrated extensively. But such similarities must be used with great caution — for instance, the Japanese, through extensive trade contacts in the last century, have absorbed much of Western culture without any significant physical change.

The Caucasoids
In the last 500 years the Caucasoid race has spread from Europe, Western Asia,

This skull, a cast of an upper palaeolithic skull from Czechoslovakia, is relatively long. Some Caucasoids, especially those in Eastern Europe, are characterized by broader, shorter heads. Most Caucasoid skulls possess a narrow nose with projecting nasal bones, and relatively small teeth, although brows may be well developed in males.

SKIN COLOUR
Skin colour is one of the main characteristics by which living races can be distinguished. It is, of course, an inherited characteristic but the mechanism of its inheritance is still not fully understood as a number of genes are involved.

As a general rule populations which live in hot, humid environments tend to have the darkest skins, since the melanin pigment appears to offer protection for skin cells against damage from too much ultra-violet light contained in sunlight. Conversely populations living in areas where there is little sunlight falling on the skin tend to have the lightest skins. This means that they are able to make maximum use of available ultra-violet light to synthesize essential vitamin-D in certain skin cells. Skin colour differences have probably partly resulted from natural selection acting on populations, balancing the need for production of vitamin-D against the problem of skin damage if too much ultra-violet light penetrates the outer layers of skin.

Even this suggestion cannot explain all the variations of skin colour found in man today. Why, for example, are there no very dark-skinned peoples in the hottest areas of the New World, and why do the Eskimos have relatively dark rather than light skins? One possible answer is that natural selection has not had time to operate on their skin colours if they only recently arrived in their present environments. In the case of the Eskimos it may even be that their skin needs protection from sunlight reflected from snow, and that they receive more than enough vitamin-D through their diet which is high in fats and fish products.

We can only make guesses about the original skin colour of early man. Most probably the first men were dark-skinned with dark hair and brown eyes and all other variations have derived from this ancestral condition by natural selection or mutation. The Cro-Magnons were apparently light-skinned, to judge by the few realistic representations in cave-art — but whether the Neanderthals had fair hair and blue eyes we shall probably never know!

Although very variable in skull form, Australoids (particularly males) are generally characterized by a long, narrow skull, prominent brows, projection of the lower face (partly related to possession of large teeth), a broad upper face, and relatively prominent nasal bones like those of Caucasoids.

India, and North Africa as far as America, southern Africa, Australia, and New Zealand. Caucasoids have variable colouring ranging from the tan skin, dark hair and eyes of the Mediterranean peoples to the very pale skin with fair or reddish hair and blue eyes of northern Europeans. They have straight or wavy hair which grows long (if allowed to!), and they also have a quite heavy body growth of hair. We cannot, of course, identify such characteristics from early human remains but the typical Caucasoid narrow and prominent nose, vertical face, (relatively) smallish teeth, and fairly robust brows (in males) can already be recognized in the Cro-Magnons of Europe and in related populations of the late Pleistocene in the Middle East and North Africa. One of the few realistic human faces painted in palaeolithic cave art shows a pale-skinned man with a prominent nose, thin lips and a dark beard. So the Caucasoids appear to have been well established 30,000 years ago in the areas where they were to be found before the great colonizations of the recent past. Caucasoids of today are little changed from their Pleistocene ancestors, the main differences being a generally less rugged build with smaller faces and teeth and broader skulls.

Australoids

Members of the Australoid race include the Australian aborigines, populations in

New Guinea and — according to some authorities — various scattered peoples of Asia such as the Veddahs of Sri Lanka (Ceylon). They tend to have robust skulls with somewhat projecting faces, large teeth, and a dark skin, often chocolate-brown in colour. Their hair is rather similar to that of Caucasoids since it tends to grow long and wavy; and some aborigines even have blond or tawny hair. The Tasmanian aborigines who died out completely in the last century were darker-skinned than the aborigines of mainland Australia and may have been more like the first human immigrants to the Australian area.

Old ideas about the arrival of man in Australia have recently had to be substantially revised. It was formerly believed that man arrived in Australia at the end of the Pleistocene period (perhaps 12,000 years ago) and this seemed to coincide with the extinction of many of the marsupial mammals originally living there. This tidy theory has now been upset by excellent evidence from artefacts and actual skeletal material which shows that man was in New Guinea and Australia over 25,000 years ago. At no time in the Pleistocene or since has there been any land connection between Australasia and South-east Asia. And so whether the first inhabitants arrived via Java and Timor, or Borneo, the Celebes and the Moluccas, they would have had to make a water crossing of 50 miles or so. This suggests that these first Australians must have

73

been the earliest known mariners. It is still not certain how many waves of immigrants came into Australia, but the earliest skeletal evidence from Lake Mungo in New South Wales (about 30,000 years old) is perfectly modern in form and like the least robust of modern aborigines. More problems in interpreting the prehistory of the Australian aborigines are posed by a collection of skeletons from Kow Swamp, not far from Lake Mungo. These skeletons reveal a population of very robust people who lived about 9,000 years ago and who are said to show some characteristics of *Homo erectus*. Some people think that this group was a remnant stock actually derived direct from *Homo erectus*, perhaps from the earlier Solo men in Java. Other scientists point out that the Australian aborigines have always been a highly variable population and suggest that the Mungo and Kow Swamp skeletons are merely extremes of a wide range which is still evident today.

Whichever view eventually turns out to be correct there seems little doubt that the Australoids were already racially distinct at about the same time as the Caucasoid Cro-Magnons were populating Europe. They may also have been more widespread at that time as suggested by remains from the Niah cave in Borneo (about 40,000 years old), and the Tabon cave in the Philippines (over 20,000 years old). Perhaps they were derived from a common stock with the Caucasoids in Asia which would explain some similarities between

them in racial characteristics. This original undifferentiated Caucasoid/Australoid stock may have also given rise to some of the populations of Asia such as the Veddahs of Sri Lanka and the Ainu of Japan who display resemblances to the Australoids and Caucasoids. The Pacific islanders and the Maoris (who arrived in New Zealand by boat about AD 1000) could also be derived from such an ancient stock, but they also show a number of signs of Mongoloid ancestry.

The Americas

Our ideas about man's arrival in America have also been greatly revised in the light of new evidence. It was assumed that there

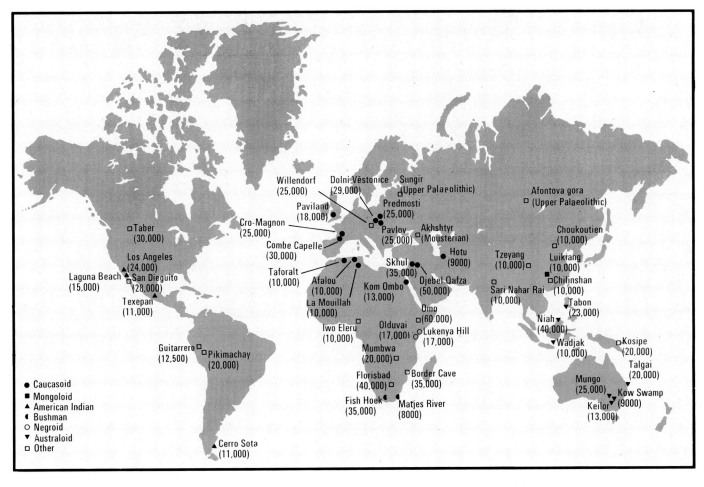

Willendorf (25,000)
Dolni Věstonice (29,000)
Sungir (Upper Palaeolithic)
Afontova gora (Upper Palaeolithic)
Predmosti (25,000)
Paviland (18,000)
Cro-Magnon (25,000)
Pavlov (25,000)
Akhshtyr (Mousterian)
Choukoutien (10,000)
Combe Capelle (30,000)
Hotu (9000)
Tzeyang (10,000)
Luikiang (10,000)
Taber (30,000)
Skhul (35,000)
Los Angeles (24,000)
Taforalt (10,000)
Afalou (10,000)
Djebel Qafza (50,000)
Sari Nahar Rai (10,000)
Chilinshan (10,000)
Laguna Beach (15,000)
San Dieguito (28,000)
Kom Ombo (13,000)
Tabon (23,000)
La Mouillah (10,000)
Texepan (11,000)
Omo (60,000)
Niah (40,000)
Iwo Eleru (10,000)
Olduvai (17,000)
Lukenya Hill (17,000)
Wadjak (10,000)
Kosipe (20,000)
Guitarrero (12,500)
Pikimachay (20,000)
Mumbwa (20,000)
Talgai (20,000)
Florisbad (40,000)
Border Cave (35,000)
Mungo (25,000)
Kow Swamp (9000)
Fish Hoek (35,000)
Matjes River (8000)
Keilor (13,000)

● Caucasoid
■ Mongoloid
▲ American Indian
◀ Bushman
○ Negroid
▼ Australoid
□ Other

▲ Cerro Sota (11,000)

too, man had arrived at the end of the Pleistocene and his arrival led to the extinction of many large mammals. But it now seems that the American Indian Mongoloids were established in North America at least 25,000 years ago and had reached the southern tip of South America by 10,000 years ago. Human remains of Pleistocene age are rare in the New World, but there are well-dated finds from Canada and the United States which are certainly 25,000 years old, and archaeological evidence suggests an even earlier date for man's arrival there. Although there is now no land contact between Asia and Alaska, in the periods of glaciation a wide land bridge known as Beringia was created by the lowering of the sea level in the Bering Straits. This land bridge allowed a considerable interchange of animals between Asia and America and early Mongoloid hunting peoples must have followed animal herds migrating across it.

The earliest American human remains are of modern *Homo sapiens* – there is no evidence so far of any Neanderthals or *Homo erectus* there – already recognizable as American Indian Mongoloids. But some archaeological evidence could put the arrival of man in America at over 40,000 years ago. So the American Indian race appears to have evolved as early as the Caucasoid race in Europe and the Australoid race. We cannot, of course, find in fossil material the typical American Indian features such as medium-brown skin colour, straight black hair sparsely distributed over the body, and the varied degree of development of an epicanthic fold (a fatty fold in the upper eyelid). But the characteristically narrow and high-bridged nose, often protruding front teeth, and a long skull can already be recognized among the Pleistocene American fossils. The frequent occurrence of the epicanthic eye-fold and certain characteristic fingerprint patterns and blood groups of present-day American Indian Mongoloids suggest that their ancestors were an early offshoot of the Asian Mongoloids and the latter continued to develop after American Indians became isolated in the New World. It was very much later (10,000

This map shows key sites in tracing the spread of the various races of man over the world, with their approximate age.

This Eskimo girl shows typical Mongoloid facial features. Her flat face has well developed eye folds, prominent cheekbones, flat nose, thin lips, coarse straight black hair, and yellow-brown skin colour.

This skull of a Greenland Eskimo shows typical Mongoloid features – a flat face with prominent sides, angulated cheek-bones, a narrow nose, and flat nasal bones. The orbits may be big and the whole-skull particularly high and straight-sided in Eskimos, occasionally being rather pointed along the top when viewed from behind.

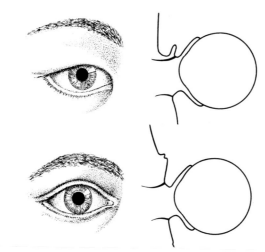

Right: The eyes of Mongoloids are characterized by an epicanthic fold in the upper eyelid (top); the lower drawing shows the Caucasian eye which has a single fatty lid.

years ago perhaps) that the Asian Mongoloids (in the form of the Eskimo) migrated into the northern areas of North America, where they live today.

Asian Mongoloids

The ancestry of the Asian Mongoloids is still unclear but their spread over Asia seems to have occurred relatively recently, and mainly after the end of the Pleistocene period. The Asian Mongoloids (also including the North American Eskimo) are typically short and stocky; their skin colour ranges from brown to almost white,

and they have straight dark hair with little body or facial growth. They often possess an epicanthic fold and their faces are flat with wide angulated cheek bones and relatively narrow noses, particularly evident in peoples like the Eskimo. Many anthropologists have considered that a flat face with fatty cheeks and relatively recessed nose and eyes are adaptations to cold conditions, insulating delicate facial structures from frost-bite, etc., while the epicanthic fold could both insulate the eye from cold and cut down snow-glare. But cold-resistance tests on Caucasoid and Mongoloid faces have not shown the Mongoloids to have any clear advantage here, and it is difficult to imagine any greater contrast than that found between the facial form of Mongoloid and Neanderthal skulls, both of which have been regarded by anthropologists as resulting from adaptation to a cold environment. So the idea that the Mongoloid race developed its characteristics by natural selection in a harsh glacial environment at the end of the last Ice Age must remain only one possible answer to the problem of its origin.

Human fossils from Asia dating from the later Pleistocene are rare, and specimens such as the Choukoutien Upper Cave skulls (from near Peking) do not look very Mongoloid – perhaps they are related to the American Indians or the Cro-Magnons instead. But some other late Pleistocene remains from China, such as the ones from Tze-Yang (Szechuan) and Liu-Kiang (Kwangsi), do show Mongoloid characteristics, perhaps taking the ancestry of the Asian Mongoloids back as far as 20,000 years ago. What is clear is that from the end of the Pleistocene the Asian Mongoloids spread rapidly down Southeast Asia, displacing the aboriginal non-Mongoloid inhabitants, although they only reached the islands of Japan (which were already occupied by the Ainu) after about 300 BC. They also spread westwards to central Asia and northwards to Siberia and Alaska (by at least 8000 years ago). Whether at a much later date (over 3000 years ago) there was contact between the Mongoloids of Asia and the American Indians of Central America remains disputed as does the problem of contacts between peoples across the Atlantic and Pacific before the great voyages of discovery made in the last few hundred years.

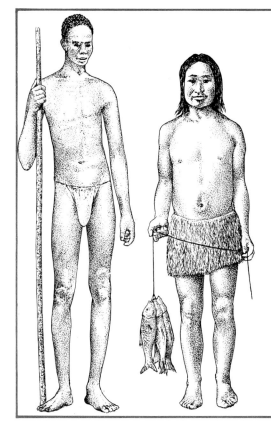

Negroid skulls show a great deal of variation but are often narrow, particularly across the base. The face is generally short and narrow, with a broad nose and delicately built cheek-bones, but the lower face may be projecting.

The Negroes

The Negroid race which now occupies much of Africa and makes up a significant part of the population in several American countries displays a wide range of variation. It includes the tallest and smallest peoples on Earth. Many of the Negroid **populations illustrate Bergmann's rule** since they are generally lightly built with limbs which are long compared with their trunk size – this gives them a proportionately large skin surface area and probably evolved to help in losing heat more easily. Their skin is usually dark, and their hair is tightly curled and black with only a sparse body growth. Negroids have especially distinctive facial features with lips which turn outwards and a projecting lower face with a broad nasal opening and delicate cheek bones. Their foreheads are generally vertical but low, with small brows compared with say, Australoids or Caucasoids. The Pygmies of central Africa are also classified as Negroids; they differ from the typical Negroids mainly in their much smaller body size. Some other mammals such as the hippopotamus have pygmy forms in forest areas and it is likely that reduced size aids mobility and survival in the dense equatorial forests. There are scattered populations in South-east Asia and the Pacific islands with superficial resemblances to the pygmies and other Negroid peoples. They are sometimes referred to as Negritoes; evidence of blood groups and fingerprint patterns suggests that the resemblances do not indicate a close relationship but are most probably the result of parallel adaptation to similar environments.

A husband and wife from the Igbirra tribe who live near Okene in Nigeria. They have typically West African Negroid features, which include the dark colour of eyes, hair, and skin, the broad flat nose, and thick everted (turned out) lips.

We might expect that their distinctive facial features and physique would make it easy to recognize Negroid skeletons among the various African human remains of the late Pleistocene and early Holocene. But in fact few which can be positively indentified as Negroid have so far been found. This may be because the Negroids developed in an area such as the Sahara or the dense tropical forests of central Africa where little has been discovered; on the other hand the Negroids may in fact have been one of the last races to develop their present racial characteristics. One fragmentary skull from Lukenya Hill in Kenya is associated with a 'Late Stone Age' industry and has a radiocarbon age of over 17,000 years BP (before the present). What there is of the cranium suggests a low robust skull by modern African standards, but nevertheless the root of the nose is flattened as in many Negroes and Bushmen of today. Other later Kenyan skeletons (probably less than 10,000 years old) such as those from Gamble's cave and Bromhead's site (Elementeita), have been regarded as evidence of Caucasoids living in East Africa, but they may in fact be early examples of the Nilotic Negroes who can be found today in areas such as the Sudan.

An indisputable Negro skeleton comes from Mali (Sahara) and suggests a tall but lightly built individual. But again it is less than 10,000 years old and this also applies to fossil material of early Negroes from Zaire (the Congo) and the Sudan. Another skeleton from the Iwo Eleru rock-shelter in Nigeria is about 10,000 years old but seems primitive (reminiscent of the Omo crania from Ethiopia, see page 70) rather than Negroid in appearance. On present evidence, then, the ancestry of the Negroids cannot be traced as far back as that of the Caucasoids, Australoids, and the American Indian Mongoloids. The spread of the Negroids into southern Africa does seem to have been a fairly recent occurrence and the earlier inhabitants of this area were apparently the peoples called Bushmen, Khoisan (from their name for themselves) or Capoid (because they live near the Cape of Good Hope).

Bushmen

The Bushmen are now restricted to limited parts of southern Africa such as the Kalahari desert, but there is evidence to suggest that they were once more widespread and could even have been the first race of modern man to evolve. The unique languages of the Bushmen and the closely related Hottentots contain clicking sounds, and this feature is also found in the language of peoples of East Africa such as the Hadza. Additionally the characteristic rock-paintings produced by the Bushmen have also been found in East Africa, and this again suggests that the Bushmen used to live in that area and perhaps even farther north.

The modern Bushmen are sometimes described as *paedomorphic* since they may retain an immature physical form into maturity. They are short in stature and have small skulls often with a pentagonal (or five-pointed) shape when looked at from above. They have smooth, vertical foreheads and a generally 'young' appearance to their faces, although there is a tendency for their skin to become wrinkled at a relatively early age. Skin colour is yellowish-brown and they often possess a kind of epicanthic fold, but these resemblances to the Mongoloids seem to be the result of parallel evolution rather than indicating a close relationship, because in general the Capoid blood groups and fingerprint patterns seem closest among modern races to the Negroes, suggesting a common ancestry with them. The hair form of the Bushmen, sometimes called 'pepper-corn', is tightly curled and tufted.

As for the ancestry of the Bushmen, there is more Late Pleistocene human material available in southern Africa than in the more northerly areas which provided so little material to cast light on the origins of the Negroid race. Some of these fossil specimens such as those from Broken Hill and Saldanha (over 50,000 years old), and Florisbad and Border Cave (probably at least 35,000 years old), seem too robust to be ancestral to such 'paedomorphic' peoples as the Bushmen – considering the time available for the transition. But some remains associated with the South African Middle and Late Stone Age (probably 20,000 years old) are reminiscent of Bushmen or Hottentot skeletons and are only slightly larger and more robust. Examples of these are the Boskop skull and a skeleton from Fish Hoek which may be 35,000 years old; there are many later specimens from sites such as Matjes River and Tsitsikamma which show that the Bushmen were well established in southern Africa by the end of the Pleistocene – about 10,000 years ago.

We can only make guesses about the spread of modern man in Africa, since large areas such as the Sahara have produced little in the way of prehistoric human remains. Some experts have suggested, in the absence of other evidence, that the Negroid peoples developed in West Africa and in the Saharan area, which has only become desert relatively recently. Some Negroid features could have originally been adaptations for life in hot open country. One group of early Negroids may have become adapted to life in the denser equatorial forests – and evolved into the Pygmies by a gradual decrease in size. Then with the coming of agriculture and metal tools the taller Negroes were able to penetrate and clear the dense equatorial forests to grow crops. Over several thousands of years the Negroes migrated into East Africa and those that took up pastoralism (cattle-keeping) continued to expand their grazing lands into southern Africa, where they gradually displaced the native Bushmen who had been occupying the area for at least 20,000 years. This process of Negro expansion was still continuing when the first Caucasoid settlers from Europe arrived in South Africa during the 17th century.

Our knowledge of the history of the races of man continues to increase as new discoveries are made and work on the biology of living races provides further evidence about the origin of 'racial' differences. Evidence about the races of man before about 5000 BC is still very incomplete for most parts of the world and, as we have seen, ideas about the arrival of man in America and Australia have had to be completely changed because of discoveries made in the last ten years. We know that great migrations of human populations must have occurred regularly in the past and continue to happen at the present time – often at the expense of the aboriginal inhabitants of the area invaded. It is not only their cultures that suffer: the Caucasoid expansion of the last 500 years reduced the numbers of aborigines or even obliterated them by new diseases brought by Europeans and by wars. We can only hope that something good can come out of past tragedies such as the extinction of the Tasmanians or the destruction of American Indian societies by this massive expansion of the Caucasoids. As the spread of 'civilized' peoples and ideas continues at an ever-increasing pace, perhaps we should realize that the so-called 'primitive' peoples have an evolutionary history as long and as interesting as our own, and that we still have much to learn from them if only we will stop to listen.

Bushman (Capoid). This cast of a fossil Bushman skull from Matjes River in South Africa shows some typical 'paedomorphic' features. It is small-faced with a flat, broad nose, has a prominent forehead, small brows, and rather bulging occipital and parietal bones. Because of the latter feature the skull may appear pentagonal (five-pointed) when viewed from above.

Amazonian Indians perform a ritual dance, their wrists decorated with bunches of leaves. Modern forms of transport are now bringing such peoples, hitherto isolated, into contact with modern civilization. Almost inevitably their traditional way of life, and their independence, will be swept away.

Stone Age Hunters

Left: Part of the shaft of an antler tool on which the head of a bison has been carved with a flint graver. It was found in Magdalenian levels in the cave of Isturitz in the Pyrenees in south-west France.

During the two million years of the Pleistocene it is possible on the one hand to trace the physical evolution of man from the hominid stage to *Homo sapiens*, and on the other to follow his cultural evolution, an aspect unknown in other animals.

Man's physical evolution is distinguished by the development of the brain. It was this that enabled him to survive against the competition of larger animals that possessed fangs and claws for hunting. Man used his brain to discover how to make weapons from stone and wood for defence and hunting. Most animals are confined by physical adaptations to one environment; man alone has been able to adapt to climates all over the world, by using his brain to invent ways of keeping his body warm through clothing and fire.

Early in the Pleistocene man began to move out of the tropics into cooler regions, leaving behind stone tools that can be used to identify and date the migrations. Many aspects of everyday life at this remote time have vanished beyond recall, but archaeologists studying remains from many Pleistocene sites have been able to undertake a certain amount of reconstruction from what remains. It is now possible to see a gradual evolution of cultural life from very simple beginnings to great technical ability, reflected in successive stone tool industries and art forms.

For long ages man lived as a hunter and foodgatherer, a way of life that even now survives in remote regions of the world. The palaeolithic ('old stone') cultures of the Pleistocene period and mesolithic ('middle stone') cultures of the Holocene (after 10,000 BC) comprise this stage. Man was then like other animals, completely dependent on his environment. Later neolithic ('new stone') cultures, from about 7000 BC in the Near East and later in other parts of the world, demonstrate man's use of his brain to think out ways of control-

ling the environment and producing his own food supply by growing crops and domesticating animals. By 3000 BC, the first complex economies based on concentrations of populations in cities can be seen in the Near East – man had reached the stage of 'civilization'.

Man takes shelter

In the warm tropical regions of Africa or Asia where hominids first evolved, there was little need to make artificial dwellings. The great apes sleep in a form of nest in the lower branches of trees, as a refuge from carnivores, and hominids may have done the same. It is probable that shelters made by early men were like structures built by hunters of recent times, such as the 19th-century Tasmanians. These tribes were perpetual wanderers, like Pleistocene men. In the summer they made a curved screen of strips of bark laid against wooden stakes, to keep the wind off them as they slept. In winter they made a 'hut' by clearing a round space in a thicket of young trees, and bringing together the tops of the trees round the space to make a roof over it. They thatched this living roof with leaves and grass.

Such simple shelters would rot away and leave no trace for archaeologists; the oldest known structures are circular settings of angular blocks of stone, found by the Leakeys at the Olduvai Gorge and dating from about 2 million years ago. These must have been arranged by the hominids and not by water action, because stones swirled about in floods become rounded.

As man moved into colder areas of the world, more substantial dwellings were needed for him to survive the winters. During the fluctuating climate of the Pleistocene, the northern hemisphere underwent four major glaciations when huge ice sheets extended almost as far south as the Alps. As each of the first three glaciations approached men retreated southwards with the warm-climate animals. But about 100,000 years ago, some time before the last glaciation, men discovered how to make fire. This discovery allowed Neanderthal man to occupy caves in Europe, and later Gravettian hunters to prey on mammoths and other animals of the freezing steppes of eastern Europe.

These hunters of the last glaciation faced extreme cold, and kept warm by

building their huts below ground, or by making the walls very thick, using stones and massive mammoth bones, to retain heat. Excavations in Czechoslovakia, Poland, and South Russia have revealed many examples of ingeniously constructed houses built about 25,000 years ago, using only stone tools and animal products. At Dolní Věstonice, in Czechoslovakia, level floors were dug into a hillside and mammoth bones and large stones were arranged to make low walls. A sloping roof, probably of animal skins, was supported by slanting posts. The diameter of one hut was 6 metres (20 feet). Much larger structures, with a diameter of 42 metres (46 yards), were found near Kiev in 1965 when a farmer digging a cellar uncovered a mass of mammoth bones. About 385 bones from at least 95 mammoths had been used to make a house. Great jawbones piled up on one another formed the walls, with massive long-bones standing on end in the ground near the doorway. Above the door was an arch formed by two curving ivory tusks, and a frame of wooden posts and bones supported the roof. In the centre of the house was a hearth with a layer of ash 20 centimetres (8 inches) thick.

Other huts were dug into the soft loose soil of South Russia. At Timonovka six substantial dwellings were found, each about 12 metres (40 feet) long and 3 metres (10 feet) wide. They had been roofed over at ground level by timbers covered with earth. One hut had a chimney made of bark covered with clay to carry off smoke from the fire. Fatty mammoth bones had been burned as fuel, and soft stone lamps gave light.

	EUROPE	AFRICA	WESTERN ASIA	FAR EAST
HOLOCENE (POSTGLACIAL)	Mesolithic cultures: Azilian, Maglemosean, etc. Microliths, flint axes, painted pebbles	Mesolithic cultures: Capsian, Wilton—microliths	Natufian cultures—microliths; 'husbanding' of plants and animals	
8000	Upper palaeolithic cultures—flint blades and burins, antler tools; cave art	Oranian blade-tools Aterian industry	Upper palaeolithic cultures	China; some blade tools among flake and pebble tools
LATE PLEISTOCENE	Homo sapiens sapiens: 32,000 BC Neanderthal man—Mousterian industries: 50,000 BC Levalloisian industries	Levalloiso-Mousterian industries	Homo sapiens sapiens—blade tools: 33,000 BC Mousterian and Levalloiso-Mousterian industries: 50,000 BC Jabrudian flake tools	Derivatives of chopper-core tools
100,000				Chopper-core tools—
MIDDLE PLEISTOCENE	Levallois technique Acheulean hand-axes Abbevillian hand-axes Clactonian industries	Levallois technique Acheulean hand-axes	Acheulean hand-axes	Malaya; Tampanian Java; Patjitanian China; Homo erectus-Choukoutienian
400,000				
EARLY PLEISTOCENE	Vértesszöllös and Homo erectus	Homo erectus—first hand-axes Australopithecus—chopper-core industries		

At Kostienki, in South Russia, builders may have appreciated that a double wall would keep out cold, just as a tent is warmer with a flysheet. The area of the floor covered with rubbish was much smaller than the total area of the floor, which suggests that, like the present-day Chukche of Siberia, the hunters built a double-walled construction with an inner compartment of skins, where they slept near the fire.

Tents and middens

By 10,000 BC, northern Europe was free from ice. Hunters moved northwards into the newly exposed areas, hunting reindeer and (as forests replaced the tundra) red deer and wild pig. These post-glacial mesolithic hunters moved from place to place with the changing seasons, leaving behind tent-stances of circles of stones. These weighted down the edges of skin tents, like Red Indian tepees. Some tribes spent summers beside lakes, hunting red deer as well as fishing and fowling. At Star Carr, in Yorkshire, a platform of brushwood covered in places with strips of birchbark was excavated from 1949 to 1951. Patches of clay had been laid down where fires were lit, using bracket fungus as tinder. Other groups lived by the sea, accumulating huge heaps of shells from the limpets, oysters, mussels, and other shellfish that they consumed in vast quantities. They lived on these rotting heaps, and even buried their dead in them. Such shell middens or 'kitchen middens' have been found round the coasts of Europe from Spain to Denmark, where a mound at Ertebølle was still occupied when the first farmers arrived in the area about 3200 BC.

Caves and rock-shelters

The remains of artificial dwellings are found only by chance – usually when trenches are dug as part of building operations – as there is no trace of them above ground. Caves, on the other hand, are

About 23,000 BC Gravettian hunters were living in the cold tundras of Eastern Europe, hunting mammoths and wild horses with wooden spears tipped with stone spearheads, or mammoth ivory lances. At Ostrava Petřkovice in Silesia, shown here, the hunters constructed stout tents with hearths in bowl-shaped hollows. Ashes from these hearths show that the hunters had discovered a nearby outcrop of coal which they used as fuel. At Dolni Věstonice in Czechoslovakia the remains of more substantial huts have been excavated, each represented by an irregular oval area of 15 by 9 metres. Inside one hut were five hearths in line, with large flat stones, possibly for seats, round them. Many stone and bone tools, and ivory ornaments, had been dropped around the hearths. Outside one hut was a working floor where a flint-knapper had made tools. Evidence for dress comes from small figurines that seem to be wearing tunics and trousers not unlike Eskimo costume.

Angus McBride

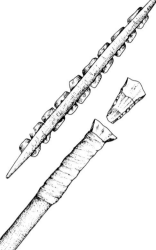

A slotted bone fish-spear with tiny flint microliths inserted as barbs. This comes from the Maglemosian culture of Denmark.

A mesolithic wooden arrow preserved in peat. The microlith used as an arrowhead is bound in place with sinew. It has a cutting edge, not a point, and is called a tranchet arrowhead.

obvious places in which to look for traces of occupation by early man. Unfortunately, investigations by antiquaries and treasure-seekers have often disturbed the ancient levels, and, in addition, many caves have been used up to modern times for animal pens, storehouses, and temporary dwellings by shepherds.

Caves form in limestone, a soft rock that is worn away through the action of groundwater. Shallow caves under overhanging rocks are called 'rock-shelters', but other caves run deep into the ground. In the Dordogne in Central France, the soft silvery limestone is riddled with caves and rock-shelters. Some contain many layers of ancient tools and animal bones, showing that they were lived in almost continuously by palaeolithic man. Others contain few layers, while many apparently habitable caves contain no occupation levels at all. It is possible that palaeolithic families lived in one cave during the winter and moved to another for the sum-

mer, choosing sites near places where the animals they hunted came to graze or to drink. By measuring the amount of sunshine, archaeologists can suggest that a cave might have been too hot to live in during the summer, but warm in winter.

Cave life had its dangers. Large rocks frequently fell from the roof, and skeletons have been found at Shanidar and elsewhere with evidence of crushing through such rockfalls. There was also no retreat from a cave if the occupants were suddenly confronted by a cave lion or bear.

Evidence from one cave at Arcy-sur-Cure showed that Neanderthal men lived there in very squalid conditions. Bones of hundreds of animals were found at the back of the cave, indicating that carcases of wild horses and cattle, reindeer, and wolves were brought to the cave and cut up, the remains being left to rot close to the tribe's living area. At this level was found the oldest known 'curio collection' – some nodules of iron pyrites of curious shape,

and fossils, shells, and madrepores, that must have caught the eye of some Neanderthal man, woman, or child. Later hunters lived in the same cave, in similar conditions; their occupation was brought to an end by an enormous fall of boulders from the roof. After this early *Homo sapiens sapiens* used the cave, levelling off the uneven floor with stones, and throwing their rubbish outside the cave. At a still later stage, the cave was occupied by Magdalenians who set mammoth limb bones upright in the floor as work-tables. The upper surfaces of these were marked by cuts from flint knives.

Keeping out the cold

The dress of Pleistocene people could have been colourful and elaborate, to judge by the costumes worn by modern primitive tribes. These are made of easily available materials – flowers, leaves, feathers, shells, animal teeth, and hair. Wood is carved into masks. Sometimes the body is tattooed or ornamented with cicatrices – scars or gashes rubbed to form large weals.

No Pleistocene body has yet been found with any form of dress preserved. Evidence can only be deduced from such sources as burials, where stone, shell, and bone ornaments can be seen in position on the body. At Barma Grande in Italy, pierced animal teeth, shells, and fish vertebrae in the neck region could be recon-

Two upper palaeolithic necklaces. The outer one, made of bone beads and pendants in which holes have been drilled for ornament, was found in a child's grave in Siberia. The inner drawing shows part of a necklace found in the grave of a young man at Mentone in France. Between pierced deer's teeth were strung two rows of fish vertebrae and one of shells.

structed into an elaborate necklace. Similar necklaces have been found at other palaeolithic and mesolithic sites. Shell and bone ornaments were apparently sewn on to a shirt, presumably of leather since there is no evidence of woven cloth in the Pleistocene. At Sungir, 210 kilometres (130 miles) from Moscow, is the grave of a man of about 55, lying on his back. Hundreds of cut and polished mammoth-ivory beads lay in continuous rows across the chest, suggesting that the shirt had no front opening. There were also two dozen mammoth-tusk bracelets in the grave. Ornaments made from the pierced teeth of arctic foxes and ivory beads were also found on the skulls of two boys buried head to head about 25,000 years ago at Sungir. These suggested that the boys had worn decorated leather caps. An ivory horse, coloured red, lay on one boy's chest. Under the chins of both were bone cloak pins.

Palaeolithic art provides little evidence of dress, since humans were not normally portrayed. But at Angles-sur-l'Anglin is a portrait of the upper part of a man with a black beard, with a reddish area below the neck that could be a leather tunic, between areas of scratches that could represent a fur cloak. The statuettes of women – 'Venuses' – give little evidence of dress.

Following food

During the Pleistocene men lived by hunting and foodgathering, moving from place to place to take advantage of seasonal foods and following herds of animals. Permanent settlement was only possible

The upper part of the skeleton of a man buried at Sungir, not far from Moscow, excavated in 1955. The grave had been lined with red ochre, which had also been sprinkled on the body and had stained the bones bright red. The man had been buried wearing clothes on which about 1500 ornaments had been sewn in rows. He wore bracelets of mammoth ivory on both arms.

A reconstruction of a deer-antler 'frontlet' found at Star Carr in Yorkshire, and dating from 7500 BC. The antlers were hollowed out for lightness, and part of the skull was drilled with holes, perhaps so that it could be attached to a leather headdress. It may have been worn for ceremonial occasions, or as a hunter's decoy.

where there was a steady food supply, such as fish or shellfish. Occupation sites are now studied in relation to their geographical background and prehistoric environment, in an attempt to work out the time of year when they would have been occupied. The range of man's diet is suggested not only by the bones of animals that have been found on living sites, but also from studies of the diet of hunters in recent times. In addition to hunting, the Tasmanians are recorded as having collected grubs from hollow trees (said to have a flavour like almonds) and to have eaten birds' eggs, snakes, and lizards, as well as shellfish. Roots were dug up with simple tools. An intoxicating liquor was made by tapping a kind of gum-tree, and allowing the slightly sweet juice to collect in a hole at the bottom of the trunk until it had started to ferment. Australian aborigines also made a variety of drinks, and smoked leaves in a hollow bamboo pipe. They chewed leaves and twigs of a certain plant as a narcotic.

Pleistocene men could also have eaten insects, enjoyed honey, and concocted drinks, but no traces of these would have survived in the ground. But on sites where pollen and other plant remains can be extracted from the soil, it is possible to establish that certain plant foods were growing in the area, and could have been used by man.

Even in the tundra of France and Central Europe during the last glaciation, Neanderthal men and *Homo sapiens sapiens* could have found plant foods. In recent times, the Chukche of north-east Siberia stored leaves and branches of willow in sealskin sacks. These were left to turn sour during the summer, and in autumn when they froze solid each sackload was cut into slices and eaten like bread with meat. Berries, too, were collected in summer, and stored in ice-pits. Meat-eaters need vegetable matter; and Magdalenian hunters of the late Pleistocene could have obtained more vitamins and iodine by eating the half-digested sour mash in the stomachs of the reindeer they hunted, as the Lapps do today.

Fish were caught by early *Homo sapiens sapiens* groups in Europe. Fishbones have been found threaded as beads, and fish are also portrayed in cave art. Later mesolithic hunters wove fish-traps from osiers,

which have survived in waterlogged conditions in Denmark. They also made fish-spears or leisters, hafting two or three deer antler points on a shaft. These were used to stab pike – and the bones of one pike with a point among them were found in Denmark. Bone was shaped into fish hooks, some so large that they could have been used inside a small fish as live bait.

Once toolmaking had been discovered, man could become a hunter rather than a food-gatherer and scavenger. On page 64 we saw that *Homo erectus* at Choukoutien ('Pekin man') hunted many animals including deer, elephants, and rhinoceroses. Animal bones which were found associated with Swanscombe Man, 250,000 years ago, included the aurochs; red, fallow, and giant deer; horse; hare; wolf; straight-tusked elephant; and rhinoceros. A study of all the animal bones found in a site at Salzgitter-Lebenstedt, in north Germany, threw light on the diet of a group of some 40 to 50 Neanderthals. They had camped for a few weeks in summer on a site in a small, sheltered valley, beside a stream. They had eaten 80 reindeer, 16 mammoths, 6 bison, 5 horses, and 2 woolly rhinoceroses. They had also caught single specimens of wolf, muskrat, a crane or swan, a duck, an extinct form of vulture, and various fish and insects. This total probably represents the game they could catch before the local resources became exhausted and they moved on.

For the upper palaeolithic, there is evidence of diet not only from animal bones found in caves and artificial dwellings, but also from paintings and engravings in caves, and other art forms. In France, hunters of the warmer phases of the last glaciation caught wild horses, cattle, and deer.

An Australian aborigine using a spear-thrower. similar to those used by Magdalenian hunters of upper palaeolithic Europe, to increase the length of the cast. A hook on the end of the spear-thrower fits into a hole at the end of the spear.

A Magdalenian 'harpoon' used for hunting reindeer. The barbed antler head was attached to the shaft by a cord. The swelling on the lower end fitted loosely into the hole in the shaft. When the spear was thrown, the barbed head would enter the body of the animal, and the shaft would become tangled in its feet, slowing it down.

A Maglemosian bone fish-hook, 7.5 centimetres long, from Denmark. The size of the hook suggests that it was used inside the bait – probably a small fish.
Below: A dug-out boat from north Holland.

This woman's head, carved with a flint graver from mammoth ivory, was found at Brassempouy in south-west France and is from the Gravettian culture, possibly 22,000 years BC. It is 3.5 centimetres high. Hair or a head-dress is indicated, but it has no mouth.

In the colder stadials, they successfully hunted mammoths, woolly rhinoceros that were even larger than the modern forms, and reindeer.

Neanderthal men and later upper palaeolithic hunters must have been both daring and skilful to hunt such formidable beasts as rhinoceros and mammoth. Armed only with wood and stone weapons, they may have organized hunts to drive animals into ambushes, over the edge of a cliff, or into a marsh. This would imply language and co-operation.

Cave art

Palaeolithic art is associated only with *Homo sapiens sapiens* between 30,000 and 10,000 BC. It took the form of paintings and engravings, sculptured friezes, and antler and bone artefacts. Art has now been found in Spain, France, Italy, Sicily, Anatolia, and in caves in the Urals. In Anatolia, similar art continued into neo-

The head of a lioness (above) and figure of a cave bear (right) are the first known objects made of clay. They were excavated from a hut at Dolní Věstonice in Czechoslovakia and date from 23,000 BC. They were preserved because they had been hardened by baking; earlier unfired efforts would have dissolved in the ground.

lithic times, while in Spain the palaeolithic tradition can be traced in paintings of an even later date, showing bronze artefacts that must be of the 12th century BC.

At one time it was thought that Eskimo art must be derived from the traditions of palaeolithic hunters from France who moved into the far north after the last glaciation, but there is now thought to be no connection. Another independent development is the art pecked into flat rock surfaces in Scandinavia in post-glacial times, showing animals, fish, and various activities. The rock-carvings of the Val Camonica, near Brescia, and other areas of northern Italy are also a separate and later development. Rock paintings are known in many parts of Africa, from the Sahara to the Bushmen paintings of the Kalahari. But there is no link between these and the palaeolithic art of Europe. Art is also found in the New World, and some Indians even put paint on the hand and made a print on the cave wall as had been done in the French caves thousands of years before; but this is simply the same form of expression, of unknown meaning, arising among peoples widely separated in space and time.

The various forms of 'cave art' can be divided into art concealed in the depths of caves, and art that was intended to be seen by daylight – small carved figures, mostly of women, but also of men and animals; decorated tools; and sculptured friezes of animals in rock-shelters.

The first manifestations of such art are associated with the Gravettian hunters and early Perigordians and Aurignacians from 30,000 to 15,000 BC. In a grave at Brno in Czechoslovakia were found the head and torso of a male figure carved in mammoth ivory. If the figure had possessed legs, it would have been 42 centimetres (17 inches) long. The head shows short hair, and deepset eyes.

A number of female figurines, or 'Venuses', have been found, including the Venuses of Willendorf, which is 11 centimetres (4½ inches) long, and Lespugue (15 centimetres; 6 inches), the former of limestone, the latter of ivory. These are fat, possibly pregnant, females, and the curves of the body are made into a design of bold stylized bulges. A number of tiny female heads have survived – two of ivory were found at Věstonice in Czechoslovakia

(4·8 centimetres; 1⅞ inches) and Brassempouy (3·5 centimetres; 1⅜ inches). Even though the Brassempouy head has no mouth, these seem more like portraits than the Venuses, which show no features. Strikingly naturalistic animals carved from ivory were found at the Vogelherd Cave in Wurttemberg in Germany, including a horse and a panther. Remarkable models found in the huts at Dolní Věstonice represent man's first (surviving) use of clay. The hunters there did not reach the stage of making clay pots, but used the clay to shape models of a mammoth, cave bear, and other animals.

The Solutreans of 23,000 to 15,000 BC cut deeply into the soft, silvery limestone of the Dordogne. At Laussel, a rock-shelter with a spring, was found the figure of a woman 44 centimetres (17½ inches) high; her head is in profile, turning towards an animal horn held in her right hand. There is a second, incomplete figure of a man, who may be drawing a bow. Two famous rock-shelters are decorated with friezes of animals; at Fourneau du Diable are reliefs of aurochs cut so deeply into the rock that they stand almost free of it, while a frieze of animals up to 1·12 metres (3 feet 7 inches) long was found at Le Roc de Sers – horse, bison, ibex, and deer, some of which were altered into other forms.

The Magdalenians of 16,000 to 10,000 BC made antler spear-throwers, up to 33 centimetres (13 inches) in length. These are of two kinds. One is carved all over in low relief with animals that may have been the desired prey. Horns and limbs are ingeniously curved round the rod-like shaft, and a forelock often served as the hook that had to fit into the end of the spear. A second type was weighted at the far end by the carving of an animal in the round. A curious group of carved and perforated antler objects are known as *bâtons de commandement*, as they were once thought to be sceptres. From traces of wear round the central hole, it is now thought that they were used to dress leather thongs.

Antler plaques engraved with animals have been found in occupation levels with domestic rubbish at the entrance of caves, where palaeolithic hunters lived. These are often identical in style with paintings found within the caves, far from daylight, and help to date them, as radiocarbon

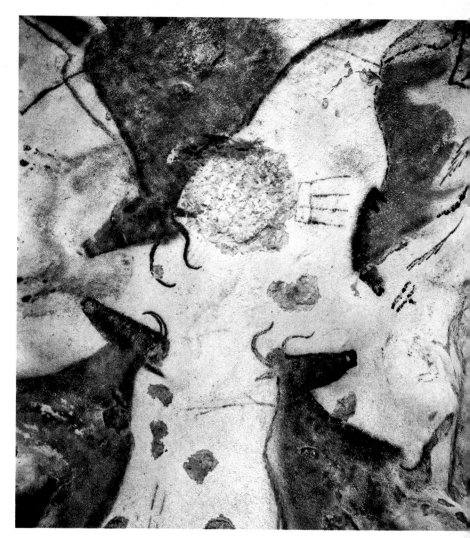

Above: Part of the ceiling of the Axial Gallery at Lascaux in the Dordogne. This is a narrow passage 20 metres long; the paintings on its ceiling, now 4 metres above the floor, were probably drawn when the floor was at a higher level. Some of the paintings are unfinished, the bodies being incompletely filled in with a colour wash. Below: A horse outlined in black paint at Le Portel skilfully gives an impression of movement.

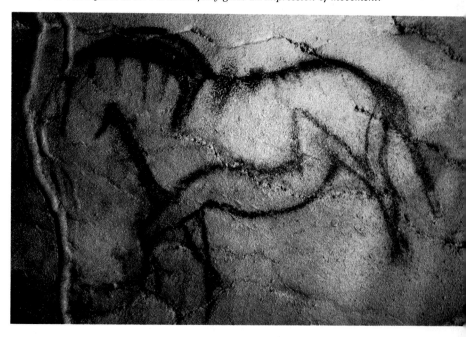

dates can be obtained from hearths in such occupation layers.

Art in darkness

Caves in limestone take many forms. Some, like Combarelles in the Dordogne, are smooth-walled tunnels. In others constricted passages lead into vast chambers, the roofs soaring into clefts beyond the beam of a torch. In some caves, the explorer has to squeeze between prickly growths of coral-like stalagmite. Graceful columns may rise from the floor, some resembling human figures or animals, while curtains and 'icicles' of stalactite festoon the roof. On the floors of some caves the footprints of cave bears and of palaeolithic men are preserved in the clay. Greasy patches on the walls show where the great bears squeezed between rocks to their lairs; striations on the walls were made when they sharpened their claws.

Palaeolithic man did not live deep in the caves. Flint implements, the bones of animals that were eaten, and charcoal from fires lie just inside the cave entrances or on the sunny terraces outside. But the paintings and engravings may lie many metres from the entrance, reached only by a tortuous journey needing artificial light, which would have had to be kept burning while the artists worked.

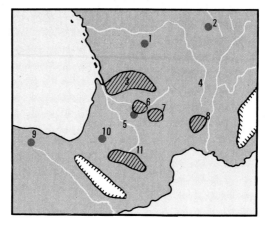

The main cave sites in south-west Europe. 1. Angles-sur-Anglin; 2. Arcy-sur-Cure; 3. Le Fourneau du Diable, Le Roc de Sers, Lascaux, and others; 4. Solutré; 5. La Magdeleine; 6. Les Combarelles, Font-de-Gaume, Les Eyzies, and others; 7. Pech-Merle and others; 8. La Baume-Latrone and others; 9. Altamira; 10. Brassempouy; 11. Les Trois Frères, Le Mas d'Azil, Le Portel, Lespugue, Niaux, and others.

Animals are represented only in profile, but on many the horns are shown as if full-face, obviously because of the difficulties of perspective. Some were outlined with a flint graver, a tool with a wide blade like a screwdriver, which could be resharpened with a well-placed blow. The walls of Combarelles form a tunnel with a mass of engravings concentrated within about 100 metres (120 yards). Most are about 0·5 to 1 metre (1·5 to 3 feet) long. Some are deeply cut; others are drawn with a fine line.

For paints, the artist used differently coloured earths from the floors. Iron oxides gave shades of red, yellow, and brown; lampblack and manganese were used for

A Magdalenian spear-thrower, 28 centimetres long, from a cave at Bruniquel in the Pyrenees. The carving of a horse shows fine observation of a leaping animal and has cuts with a flint graver to indicate the mane and shaggy coat. The perforation was probably for suspension from a thong.

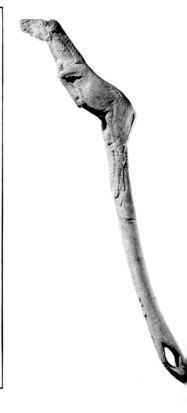

ABBE HENRI BREUIL

The Abbé Breuil's drawing of a jumble of bison and a 'bison-man' at the cave of Les Trois Frères.

From the beginning of the 20th century until his death in 1961, one man, the Abbé Henri Breuil, dominated the study of cave art. Even in the 1950s the elderly priest was called upon to make the undignified crawl through narrow caves to pronounce on the authenticity of newly discovered paintings.

Breuil's career as a defender of paleolithic art began in 1896 when he saw copies of paintings from the cave of Pair-non-Pair in the Gironde, France. They had been partly covered by undisturbed soil containing evidence of human habitation. Breuil reached the obvious conclusion — that the paintings were older than the deposits of soil, and so could not be recent fakes, and in 1901 published a paper stating this. In the same year more art caves were discovered in France, and Breuil was one of a party sent by a learned society to examine them.

He devoted the rest of his life to the work of studying and recording cave art. Many of the familiar reproductions of cave-paintings are in fact accurate drawings made by Breuil; this was the only way they could be copied before colour photography. But even today his drawings are of immense value; they often record paintings that have since decayed. Moreover, a human artist can often make sense of what the camera sees as a jumble of lines — especially where conditions for photography are far from perfect.

blacks. The paints were either used as crayons, or pounded in a stone mortar and mixed with water or urine; paints mixed with grease would not have adhered to the damp walls of a cave. The shoulder blade of an animal possibly served as a palette. Paint was stored in hollow bone tubes, and applied to the rock with some form of brush – perhaps the tip of an animal's tail, a bunch of hairs tied to a stick, or even a frayed twig. These would give an outline. For a wash, pads of moss or leather or perhaps a hare's foot might have been used. The mammoth at La Baume Latrone must have been painted with the fingers, as parallel lines form its body and legs.

Hands were sometimes dipped into paint and pressed on the wall. These give 'positive' prints. Others are 'negative', and archaeologists have made experiments to find how these were produced. A cloud of paint, fading away at the edges, appears round a negative hand. Paint must have been blown round the hand as it was held

Above: Part of the frieze from the Hall of Bulls at Lascaux, a comparatively small cave which contains some of the finest prehistoric art ever found. Unfortunately exposure to light and to the passage of thousands of tourists caused the paintings to deteriorate and the cave has been closed to visitors.

These aurochs cows, carved in deep relief, were found on a block of stone at the entrance of the Fourneau du Diable cave in the Dordogne region of France. They date from Solutrean times. The block was one of a number found at the entrance of the cave, in the area where people lived.

This bison from the ceiling of the Painted Hall at Altamira in northern Spain has been skilfully positioned round a projecting boss of rock. Using a red wash for the body, the artist has drawn the legs, horns, and tail in black; the surface of the rock shows around the limbs as shading.

against the rock – perhaps as dry powder on a slab of stone or bone.

Many paintings and engravings can be proved to be Pleistocene in age, since layers of soil containing palaeolithic tools have accumulated – in some cases over the actual art, in others blocking the entrance to the decorated areas. Engravings at Combarelles, and elsewhere, are covered with a layer of stalagmite that must have taken thousands of years to form. Further evidence exists in the discovery in occupation layers of bone plaques engraved with animal heads in exactly the same style as the animals on the cave walls.

What was their purpose?

Many of the paintings and engravings on the cave walls are beautiful representations of animals that could have been enjoyed as works of art. But there are many arguments that suggest that there was more to them than 'art for art's sake'.

First, apart from the sculptured friezes at rock shelters, the art did not decorate the part of caves where palaeolithic man lived. To visit the paintings, as well as to paint them, involved difficult and even dangerous journeys into the depths of caves. And paintings are not all found at eye-level. Some are on the ceiling, others near the ground or in crevices where they

can hardly be seen, and where the artist must have had to crouch most uncomfortably to execute them. Some animals are also drawn almost on end and many are superimposed, even where there was room on the wall to paint on a fresh area. This suggests that it was the act of portrayal that was important, rather than making a work of art. At Les Trois Frères, in the Pyrenees, are panels that are a confused mass of engraved lines, from which one can painstakingly pick out outlines of animals of different sizes and even curious, perhaps masked, human figures.

It has been suggested that the motive behind the art was 'hunting magic'. Hunters of recent times have been recorded in many parts of the world as performing a pre-hunt ritual of drawing an animal and 'killing' it by stabbing the drawing, in the belief that this will bring them luck in the hunt. Lines like arrows drawn on one of the horses at Lascaux, and what seem to be projectiles painted close to a number of animals, support this theory.

A recent theory claims that the art represents a fertility cult, with certain animals as symbols of male and female – the horse representing the male element, and bison and aurochs the female. It is rather difficult to accept this theory, as many different animals are found side by

Amber, a fossil resin found on the shores of the Baltic, was prized in prehistoric times and used for trade. The cave bear (top) and elk head were made by the Maglemosians of Denmark.

The Magdalenian culture of the cave area of the south of France and northern Spain was succeeded by that of the Azilians. Their only contribution to art is numerous small painted pebbles like these, dating from about 8000 BC. The designs, painted in red ochre, may possibly be highly stylized human figures; the purpose of the stones is unknown.

side or overlapping; and no sexual parts are indicated on animals by palaeolithic artists, which is curious if they were practising a fertility cult. It is also suggested that the caves might have been shrines, decorated as an entity and not piecemeal as the 'hunting magic' theory suggests. This does seem a valid point, certainly in regard to the Main Hall of Lascaux with its procession of great animals round the walls, and the painted ceiling at Altamira.

Curiously enough, few human figures are portrayed in cave art. The portrait at Angles-sur-l'Anglin, of a man's head in profile, is unique. Otherwise, there are only scratched caricatures of faces, or of human figures that seem to be masked or draped in animal skins, as at Trois Frères. These, unlike the animals, are shown with male sex organs. Perhaps the most famous is the Trois Frères Sorcerer, a horned figure only 73 centimetres (29 inches) long, high on the wall of a cavern deep in the Pyrenees. These portrayals are regarded as possibly 'medicine men' who carried out rituals connected with the paintings; but they are very crudely executed compared with the care expended on the naturalistic animal representations.

Many forms of 'signs' are found, both painted and engraved, on cave walls. Some look like projectiles, but others are box-shaped, or axe-shaped, or lines of dots. A red axe-like sign was painted at the end of a decorated area at Trois Frères, as if to say 'go back', and a line of dots in an undecorated passage leading to the decorated area could indicate that the visitor was going in the right direction – but they could also be where an artist was trying out his brush.

While it is not possible to look into the minds of palaeolithic artists, it may be that at least some caves were shrines, presided over by masked priests. To these hunters went to sacrifice to the spirits of animals represented on the walls; or groups of youths were led for endurance tests or initiation rites before being accepted as adults and warriors. Such rites are known to have been practised in different forms in many parts of the world.

Whatever the purpose of the palaeolithic art, in Europe it grew out of a way of life based on the hunting of the reindeer, mammoth, and other animals of arctic tundra conditions. With the coming of a warmer climate at the end of the Pleistocene, the environment gradually changed, and art took different forms. In the south of France, hunters of the Azilian culture painted geometric designs on pebbles in red ochre. In northern Europe, mesolithic hunters made antler artefacts like their Magdalenian forbears, but did not carve them with the same naturalistic animal representations. Tools were decorated with geometric incisions, scratched human figures, or with patterns of smooth pits applied with a drill. In Denmark, now free from ice, lumps of amber were carved into animal forms.

In Spain and Anatolia there were far less dramatic climatic changes, and the tradition of painting caves and rock shelters continued into later times.

Part of a group of thirteen figures engraved in full daylight on a wall of the cave of Addaura in Sicily, dating from about 8000 BC. Each figure is about 25 to 38 centimetres high. The drawings are remarkably naturalistic, but curiously their hands and feet are not shown. Some of the men may be masked. The scene may show an initiation or a ritual; the two central figures may possibly show bound men being executed by being thrown over a cliff.

Man the Toolmaker

Man has neither claws nor fangs to pierce the skin of animals, and the hard husks and rinds of vegetables and fruit. Digging up grubs and roots is difficult in hard, dry soil. Even at the hominid stage, hunger must have stimulated a hunter to pick up a sharp stone, a stick, or the broken bone or horn of a dead animal. A tool *using* stage must have preceded tool *making*.

Once tools began to be shaped, the techniques would be passed on by example and imitation. Four basic traditions of toolmaking can be distinguished during the Pleistocene. The oldest are the *chopper-core* industries associated with *Australopithecus*, *Homo habilis*, and *Homo erectus* in the Early and Middle Pleistocene. The *Acheulean hand-axe* industries are associated with *Homo erectus* and early Neanderthals, beginning in the Early and persisting into the Late Pleistocene. *Mousterian* industries of the Late Pleistocene are associated with *Homo sapiens neanderthalensis*. Upper palaeolithic blade industries are associated with *Homo sapiens sapiens* of the Late Pleistocene.

Chopper-core cultures
The oldest-known tools are pebbles from which flakes have been struck to make a sharp cutting edge. These core tools were used for chopping meat or shaping wood.

A pebble-tool from Olduvai.

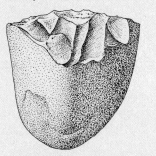

Below: A greenstone boulder flaked by Pekin man to make a chopper.

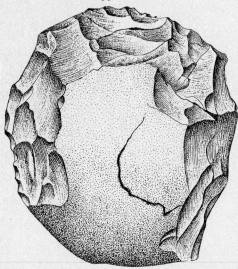

At Olduvai, such tools are called the *Oldowan* culture, and are associated with the remains of early hominids of about 1,750,000 years ago. Such tools have also been found in South Africa, where chopper-core tools are found with Australopithecines at Sterkfontein Extension Site and Swartkrans. Hominids making these tools seem to have moved from East Africa into the Near East and Pakistan, and farther east they are found in Burma, Malaya, Java, Borneo, Celebes, the Philippines, Indo-China, and north China. At Choukoutien near Peking chopper-core tools were made of quartz, sandstone, and rock crystal. They were very crudely shaped, and the waste flakes were utilized with little or no secondary working. These tools were associated with the remains of Pekin Man, *Homo erectus*, dating from about 400,000 years ago. *Homo erectus* remains were also found with chopper tools at Vértesszöllös, Hungary, dated to 400,000 years ago, but no human bones have yet been found associated with Clactonian tools in southern England and north France, of about 350,000 years ago. In Europe, chopper-core industries were replaced by other traditions; but in the Far East the old tradition continued into the Late Pleistocene. Choppers made by the same technique as that used by Pekin Man were found in the Upper Cave at Choukoutien; these deposits were of Pleistocene date, and in them were skulls of *Homo sapiens sapiens* – modern man.

Acheulean hand-axe cultures
During the Lower Pleistocene, somewhere in central or northern Africa, chopper-core tools developed into bifacially worked pear-shaped tools. Those with a point are described as *hand-axes*, those with a cutting edge instead of a point as *cleavers*. Experiments have shown that they are as good as a modern steel knife for skinning animals. They were all-purpose tools. Hand-axes had a sharp point for digging, an edge for cutting, and a butt for hammering.

Acheulean hand-axe cultures take their name from the French site where they were first found, but we now know that they first developed in Africa. Sites in Morocco, at Olduvai, and elsewhere show a long series of levels in which chipped pebbles can be seen developing into true hand-axes, the technique of shaping them changing from blows with a stone to delicate shaping with a wooden bar or bone.

Hand-axes are found all over Africa, from the Cape to the north, where they have been found with *Homo erectus* jaws at Ternifine in Algeria and Casablanca and Temara in Morocco. This evidence suggests that hand-axes were made by *Homo erectus*, during his Lower Pleistocene evolution in Africa from the Australopithecine stage.

During warm interglacial periods, Acheulean hand-axe makers moved north from Africa into Europe. At Swanscombe, Kent, beautifully shaped hand-axes have been found above Clactonian chopper tools. Associated with the hand-axes were the parts of the incomplete Swanscombe skull, a form of man evolving towards *Homo sapiens*. The hand-axe industries evolved in warm

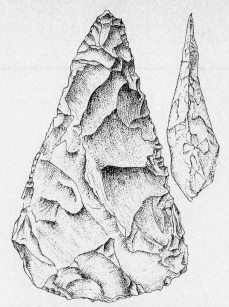

An Acheulian hand-axe (side view inset) like that found with the Swanscombe skull.

environments; their tools are found in southern Britain, north France, and south of a line from the Rhine to the Danube. Hand-axe cultures in the Near East suggest a movement out of Africa in the Middle Pleistocene, and a movement into central and southern India (*Madrasian culture*). Indian hand-axes and cleavers are exactly similar in tool technique to African examples, and must represent an immigration movement or spread of technique. In Africa, the Acheulean cultures were well-adapted to the warm forest environment, and persisted into late Pleistocene times.

Mousterian cultures
Mousterian tools are generally found with remains of *Homo sapiens neanderthalensis*. The tools suggest an evolution from Clactonian chopper-core industries, as they are made from thick flakes shaped by fine secondary working into side-scrapers (sharpened along one side), knives, and typical triangular points, that may have been hafted as spearheads. Industries from sites in the maritime lowlands of western Europe include small hand-axes, showing influence from Acheulean peoples and probably reflecting the availability of wood. Mousterian industries in the interior of France lack hand-axes, and these areas would not have been forested in the conditions of the last glaciation, when Neanderthal man occupied Europe. Mousterian tools are found at sites all round the Mediterranean Sea, and from France eastwards to Shanidar, in Iraq.

Mousterian tools – a side-scraper (left) and a point.

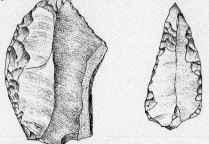

Upper palaeolithic blade industries

About 34,000 years ago, a new technique in toolmaking is found in industries associated with bones of *Homo sapiens sapiens* in the Middle East, including the Kara Kamar cave in Afghanistan and sites in Iran and Iraq. This was the striking of long, thin blades from a prepared core, using a hammer and punch. These blades were then chipped delicately into many different tools such as knives, awls, and scrapers. This *Aurignacian* culture spread with the first *Homo sapiens sapiens* into Europe about 32,000 years ago, and is found in caves as far west as France in levels above those containing the Mousterian tools of Neanderthal man.

Different forms of blade tools found superimposed in caves show migrations and regional developments of different groups of *Homo sapiens sapiens*. In Hungary and Czechoslovakia the *Szeletian* culture of about 27,000 years ago is distinguished by bifacial leaf-shaped spearpoints. The *Gravettian* culture developed in South Russia about 25,000 years ago and is characterized by narrow knife blades.

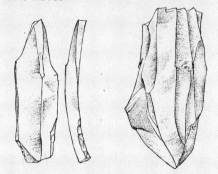

Two views of a Magdalenian graver (left), and a Magdalenian blade-core.

Solutrean artefacts are distinguished by the use of very fine flaking to make excessively thin spearpoints up to 18 centimetres (7 inches) long. It is now considered that there was no group of tribes with a Solutrean culture; but that this technique spread west from Hungary to France between 19,000 and 15,000 years ago.

About 15,000 years ago, the *Magdalenian* culture, with its art, arose in France. Flint tools included many forms of graver or burin, for engraving outlines of animals on cave walls and for carving antlers into barbed points, 'batons', and spear throwers. The Magdalenian culture spread from France into northern Spain, Belgium, Switzerland, south Germany, and Czechoslovakia.

After 10,000 BC, the melting of the ice sheets and increased warmth allowed trees to grow again in northern Europe. Wood was now more easily available for tools and weapons, and *postglacial* or *mesolithic* cultures are distinguished by the use of microliths – tiny flints that were used to tip arrows, or were set in wooden hafts. Microlithic industries are found not only in Europe, but also in Africa and Asia.

Making flint tools

Making tools from flint is skilled work, but several archaeologists have taught themselves how to produce tools like those of Pleistocene man.

Striking a block of flint in the centre is useless; the block shatters into unwieldy pieces. The only method is to detach flakes from the edges of the block. Once the correct angle at which to strike the flint is found, much less force is needed to detach a flake than to break up the whole lump by a central blow.

When a flint is struck on the upper surface, a flake falls off the lower surface. After one or two flakes have been removed by striking at the same place, the side of the block becomes too steep, and the block has to be turned. The point of impact is called the 'striking platform'. Chipped flints demonstrate that the place from which a flake was removed was used as the striking platform to remove another.

A curious pattern, resembling a mussel-shell, is produced on a struck flake, which will break off in what is called 'conchoidal' ('shell') fracture. The force of the blow travelling through the stone produces a raised cone formed immediately below the striking platform. This runs into a 'bulb of percussion', often with a tiny flake detached that leaves a 'bulbar scar'. Raised ripples show the tearing effect of the blow and the direction in which it was struck.

Stones are often broken by natural forces. The sea, or waterfalls, can dash stones against one another, even producing bulbs of percussion. Only man-made tools show purposeful turning of the stone to detach flakes from more than one direction.

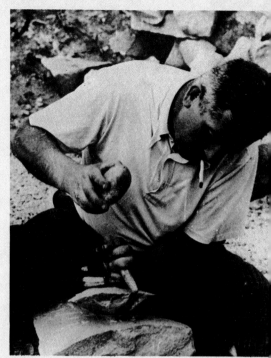

Making a stone tool by hammering a flake from a core with a piece of wood.

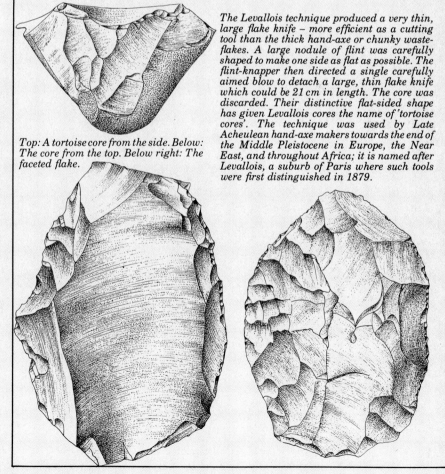

Top: A tortoise core from the side. Below: The core from the top. Below right: The faceted flake.

The Levallois technique produced a very thin, large flake knife – more efficient as a cutting tool than the thick hand-axe or chunky waste-flakes. A large nodule of flint was carefully shaped to make one side as flat as possible. The flint-knapper then directed a single carefully aimed blow to detach a large, thin flake knife which could be 21 cm in length. The core was discarded. Their distinctive flat-sided shape has given Levallois cores the name of 'tortoise cores'. The technique was used by Late Acheulean hand-axe makers towards the end of the Middle Pleistocene in Europe, the Near East, and throughout Africa; it is named after Levallois, a suburb of Paris where such tools were first distinguished in 1879.

The Farming Revolution

During the Pleistocene and early Holocene men obtained their food by hunting, fishing, and fowling, and by collecting plant food – seeds, berries, nuts, and shellfish. This is a food-gathering economy; and in some parts of the world tribes of food gatherers have survived into recent times with a way of life that has hardly altered over thousands of years. People who live in this way are rarely able to settle for long; hunters and food gatherers, today as in the time of Neanderthal man, set up camp in a cave or in the open, but must leave when the game in the area is exhausted or moves away.

A settled life can only be achieved where there is a permanent supply of food, which rarely occurs in nature. The Indians of British Columbia could live in settled villages since their way of life was based on salmon fishing. A similar way of life existed at Lepenski Vir on the Danube in what is now north-east Yugoslavia. This was a mesolithic village, occupied from 5400 to 4600 BC, again with an economy based on fishing. But these were exceptions to the general nomadic pattern of life of hunter-gatherers.

In a nomadic society the search for food often becomes a full-time occupation. The cave-dwellers of palaeolithic Europe had had comparatively little difficulty, since there were plenty of fish in the rivers and large herds of animals to hunt. As a result they had time to develop their magnificent cave art. But with the end of the last glaciation the environment gradually changed. Forest replaced tundra in Europe; but in the Near East deserts began to form as rain-bearing winds retreated north with the ice-sheets. The climatic change is reflected in the bones of animals found at sites occupied by man, the woodland fallow deer of palaeolithic levels being replaced in mesolithic levels by gazelle, which can tolerate drier conditions.

The neolithic Linear Pottery people were the first farmers to cultivate the wooded areas of Central Europe. They settled in one place for about ten years, clearing trees for fields, and moved to another site when the soil became exhausted. Many years later, when the soil had recovered, the site could be reoccupied. At Bylany in Czechoslovakia (below) there were no less than 21 habitation levels. The village consists of five or six long wooden houses, six metres wide and on average 20 metres long, built of oak posts with hurdles lashed between. These were then plastered with clay and then decorated with ochre. The clay pit was later used for rubbish. On the left an old man is seen making shell bracelets; on the right, a woman grinds corn into flour to make bread in one of the many ovens found here.

A change from well-watered forest environment towards open plains and seasonal droughts, though gradual, would tax the ingenuity of food gatherers. But people learned to adapt to changing conditions, first by husbanding existing crops such as wild grains and animals like wild sheep and gazelle, and then by cultivating and domesticating them. And this sort of farming, with a constant and static food supply, made possible the growth of permanent settlements.

A settled existence inevitably brought with it immense cultural changes. People who lived a nomadic existence would try to travel light; what food they needed to take with them would be carried in skin containers, gourds, or rough baskets. Pottery of unfired clay was too fragile to be practical. But when people settled down they were able to make large vessels of stone, and pottery was developed. The cultivation of plants led to spinning and weaving vegetable fibres and then wool from the recently domesticated sheep and goats. Eventually specialized crafts developed, villages grew into towns, and urban civilizations emerged.

Early food production

The knowledge of food production was acquired at different times in different parts of the world. In the Near East, far-mers were living in permanent villages by 7000 BC. In the Far East the oldest villages are those of the Chinese neolithic Yang-shao culture in the fifth millenium; and the earliest farming and village life in the New World began in the valley of Mexico about 1500 BC.

Archaeologists are searching for sites that will throw light on the very first stages of food production in each area. Such sites are difficult to identify, for there was no sudden change from food gathering to food production. There are still economies today that are transitional between these stages; the Lapps exercise control over herds of reindeer, but have not domesticated them. The herds wander at will, followed by a group of Lapps who 'own' a herd although they do not control the breeding or confine them in any way. Such a husbanding stage of control may have been practised by upper palaeolithic peoples such as the Magdalenians, whose economy was similarly based on the reindeer. The Natufians of Palestine, about

Above: A sandstone boulder carved into a human head with a fish-like mouth, found at Lepenski Vir in Yugoslavia. It is 40 centimetres high. It may represent a god worshipped by the mesolithic fishermen living there between 5400 and 4600 BC.

95

Moroccan goats browse in the trees. The agility of goats, allowing them to climb up to high branches, has contributed greatly to the deforestation of large areas of the Near East and Europe since their domestication by neolithic farmers. As well as damaging standing trees, goats eat the shoots of saplings and so prevent a tree cover establishing itself again.

ARCHAEOLOGICAL EVIDENCE
Archaeologists can identify the animals eaten by the inhabitants of a site by the bones they excavate. Traces of plants collected or cultivated can also be found, although plant remains are less likely than bones to survive. But sometimes a house burned down, and any grain stored in it may be found carbonized – blackened, but identifiable as wild or cultivated wheat or barley. Plant remains may also be preserved in peat, and a method has recently been devised of extracting plant fragments from dry soil during excavations. Corn grains, apple pips, and many types of seeds have been identified from clear impressions on baked clay, formed when seeds were pressed into the side of a pot or brick as it was being shaped.

The fields where ancient crops were grown rarely leave traces – they have been overlaid by later deposits, removed by erosion, or obliterated by later cultivation. But in parts of Britain and Denmark field-systems of the first millenium BC can still be seen, defined by *lynchets* – low banks formed by soil-creep during centuries of ploughing.

Many of the implements used by ancient farmers were of wood and have disappeared without trace, except in some waterlogged deposits. The oldest known ploughs to have survived are those from peat-bogs in Denmark, dating from the last centuries BC. Before the invention of the plough the soil was broken up with wooden digging-sticks, or hoes of wood or of deer antler. Baskets were used for sowing, harvesting, and winnowing. The oldest to survive was found carbonized at Çatal Hüyük in Anatolia, dating from the seventh millenium BC. A fifth-millenium basket containing grain was found in the Fayum, Egypt, in a basket-lined storage pit; it was preserved by desiccation in the hot, dry sand.

Food-gatherers collecting the seeds of wild cereals were the first to make the sickle – a curved piece of wood set with flints. Other reaping knives were straight pieces of bone with flint inserts. Experiments have shown that 2 kilograms (4 pounds) of grain could be harvested in an hour using a flint sickle.

9000 BC, hunted gazelle to the exclusion of other animals and may have husbanded them. So it is possible that before man exercised complete control over animals in full domestication, there was a stage where he selected and conserved a particular species.

Domesticating animals

It is probable that animals and grain were domesticated at about the same time. When cereals were first cultivated, animals may have become attracted to man as a food provider; there is evidence of mixed farming at Jarmo, where stubble left after harvesting would have been eaten by animals whose manure would then have enriched the land. Animals were initially domesticated for their meat and hides, but as they became more docile they would have been milked and eventually valued as pack animals.

Finding conclusive evidence of animal domestication is not easy. The bones of sheep found in levels dated to 9000 BC at Zawi Chemi Shanidar in Iraq have been claimed as the remains of the oldest domesticated animals, largely on the basis that 60 per cent of the bones were those of yearling animals. This suggests that the animals were penned and the young removed and eaten while still tender. Following comparisons of animal bones from Neolithic sites with those of wild animals

from palaeolithic sites, smaller bones are sometimes taken as a sign of domestication, as in the case of pigs found at Çayönü, Anatolia (7000 BC).

In the Near East goats and sheep were the earliest animals to be domesticated (dated to 7000 BC) and their bones are found on many neolithic sites. Pigs were probably domesticated in different areas independently. Using the criterion of smaller bones, the oldest domesticated cattle were those at Argissa in Greece (6500 BC); in the Near East the oldest are those found at Ali Kosh in Iran (5500 BC).

The earliest evidence for the domestication of the dog comes from the Jaguar Cave

Weighted digging sticks like this were probably used to cultivate the soil.

in Idaho, USA, which was associated with a hunting culture of about 8400 BC. The horse was domesticated from the wild tarpan in the steppes of south Russia, probably before 4000 BC.

Domesticating grain

Plants may have been domesticated in similar stages. The 'noble grasses', the ancestors of wheat and barley, grow wild in many parts of the Near East, sometimes forming huge stands, and would have provided an ample supply for food gatherers. But men showed their ability to shape their environment by growing grain where they wanted it, and permanent communities grew up.

It seems likely that grain cultivation started accidentally. As bundles of wild cereals were being brought to the settlement, seeds would have dropped out. In time a crop would appear. Only later, when this phenomenon was understood, would grain cultivation have been deliberately carried out.

The occurrence of wild and cultivated grain together indicates that domestication was just beginning. This is the case at Ali Kosh in Iran (7000 BC) and Jarmo in Iraq (6750 BC) where cultivated and wild einkorn (a one-grained wheat) have been found. Wheat and barley were the most useful grains to cultivate since they were very nutritious and easily stored. Emmer wheat, the most widespread of early cultivated cereals in the Near East, has been found in levels of about 7000 BC – at Beidha in Palestine, Haçilar and Çayönü in Anatolia, Jarmo in Iraq, and at Ali Kosh in Iran. Domesticated barley has been found at Jarmo and Ali Kosh, dating back to the same levels, and barley may well have come into cultivation as a weed in the wheat fields.

Below: The painting of pottery began in the early sixth millenium BC in the Near East. Many beautiful painted pots have been found in Anatolia, decorated in paint made from red ochre on a pale buff surface. Patterns were abstract and geometric (chevrons and lozenge patterns) but also derived from nature, such as stylized bulls' horns. This hollow vessel in human form may have been used for rituals, or may represent an idol. The eyes are inlaid obsidian.

POTTERY

As early as 23,000 BC the palaeolithic hunters of Dolní Věstonice used clay to model animals, but hunters and food gatherers could not carry heavy and fragile pots on their wanderings. Pottery was not made until village life developed during the seventh millenium BC in the Near East, when farmers needed clay vessels to store grain and for cooking. Probably the first pots were unfired, so would not have survived. The oldest known traces of pottery come from Ganj Dareh, Iran, and have been dated to before 7000 BC. They were hardened by accidental burning.

Early pots were made by coiling a long roll of clay above a modelled base. The sides were then smoothed to obliterate the joins between the coils. After drying in the sun, pots were fired, first in a bonfire and later in a kiln.

Early potters discovered by experiment how to mix particles of stone, shell, or straw with the clay (modern potters use 'grog' of pounded sherds) to help prevent shrinkage and cracking during firing. A coarse fabric could be hidden beneath a 'slip' of fine clay into which the pot was dipped before firing. Potters appreciated the porosity of a clay pot for keeping water cool in a hot climate, but they also learned to burnish the surface of a pot with a smooth stone, so that it would be less porous. They learned how to produce dark or pale pottery by controlling the access of air during firing.

In the Near East kilns with a perforated floor must have been in use by 5500 BC, although none has been found. The beautiful painted buff pots and hollow idols of chalcolithic Haçilar (Anatolia) and of the Ubaid and Samarra cultures of Iraq could only have been produced in such kilns, protected from the dirt of the fire. The use of potters' marks on many Samarran pots suggests professional potters.

The oldest form of potter's wheel was a flat disc pivoting on a rounded stone, turned by an

A painted limestone statuette of an ancient Egyptian potter at his wheel, from the tomb of Ne-inpu-Kau at Giza, c. 2400 BC. With one hand the undernourished potter turns the disc, with the other he shapes the pot.

assistant. This was in use during the fourth millenium BC in the Near East, and before 2000 BC in Crete, where examples have survived. The use of the potter's wheel spread through western Europe only in the first millenium BC, reaching Britain about 100 BC.

Early peoples each had distinctive traditions of potting, their vessels varying in shape and in decoration. Shapes were often copied from containers made from gourds, wood, leather, or basketwork, all traces of which have vanished. Many pots were copies of bronze or silver vessels that were too expensive to be in general use.

ROM–G

Other plants than grain were domesticated; in tropical climates 'root' crops of yams, manioc, and potatoes provided a plentiful food supply with little effort. But in areas where farming was based on seed crops there was a constant stimulus to improve the yield and find hardier varieties; successful mutants and hybrids which were larger and easier to harvest would be selected for sowing, and men learned to improve crops by hoeing and ploughing, and through irrigation. While the root farmers have not progressed beyond a basic level of food production, the seed-growing farmers have developed the world's great civilizations.

Near Eastern developments

The first stage towards a farming economy is represented by sites of 9000 to 7000 BC, where food gatherers were husbanding wild animals and wild cereals. This was a time when the climate was slowly warming, so outdoor sites could be occupied as well as caves. The oldest sites of this stage are Karim Shahir in Iraq, Zawi Chemi Shanidar also in Iraq, with possibly domesticated sheep, and Natufian sites in Palestine with possibly domesticated gazelle. Food collection now intensified, and flint blades have glossy edges (sickle sheen) which show that wild cereals were being reaped. Numbers of grinding stones

are found, probably for crushing ochres; but they could have ground seeds of grasses and later of cultivated cereals.

Husbanding cereals made it possible for people to live together in villages, at least seasonally. The lowest levels at Ganj Dareh in Iran have been dated to 8400 BC. Clay figurines were made at that time, although the first clay pots from the site were not made for another 1400 years. At Mureybit in northern Syria are traces of rectangular structures from the early eighth millenium, and there were mudbrick houses with several rooms by 7500 BC. Cereal grains identified as wild einkorn wheat have been identified there.

By about 7000 BC farmers were living in permanently occupied villages and growing domesticated varieties of wheat and barley. These early sites were in areas where there was enough rainfall for agriculture without irrigation, which did not develop for about another thousand years. At Jericho in Palestine was a settlement of about 4 hectares (10 acres) in extent, surrounded by a rock-cut ditch and towers up to 9·1 metres (30 feet) high. Possibly some 3000 people lived there, in mudbrick houses which had several rooms. Rush mats covered the floors which were of burnished pink or cream plaster. There was as yet no pottery; vessels were carved out of limestone, and such perish-

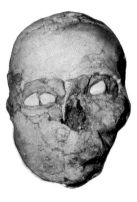

A human skull with surface features modelled in plaster and pieces of shell inlaid for eyes, found at Jericho and dating from the early seventh millenium BC. Some primitive peoples have believed that the soul resided in the skull, and have collected heads either to have power over the souls of their enemies, or for ancestor-worship. Groups of plastered skulls were found in neolithic Jericho, not as a feature of burial but evidently with some ritual significance.

A Berber woman spinning in a way unchanged since neolithic times. The weighted spindle revolves and twists the wool fibres into a continuous thread. The woman pulls out the fibres carefully to make the thread of a constant thickness.

THE FIRST WEAVERS

Traces of ancient cloth and baskets have rarely survived, but the hot dry climate of Egypt has preserved baskets of the fifth millenium BC, and the oldest examples of woven cloth there date to the fourth millenium BC. A length of cloth of the early sixth millenium BC was found inside a human skull at Çatal Hüyük, Anatolia, preserved through accidental charring. Complete costumes of men and women of about 1200 BC have been found in Denmark, preserved by waterlogging when rain water seeped into oak tree-trunk coffins. Cloth-wrapped axes were sometimes buried with warriors in second-millenium Britain; fibres were sometimes replaced with metal oxide, so the weave of the vanished cloth can be studied.

Even where specimens have not survived, archaeologists can infer that weaving was practised from finds of artefacts used in spinning and weaving. Small perforated cylindrical or triangular lumps of clay or stone are loom-weights that were attached to the vertical warp threads on an upright loom, all traces of which may otherwise have vanished. Bone combs that may have been used to pluck wool from sheep or to press the weft threads together have been found in Britain from about 500 BC. More evidence comes from impressions of cloth found on clay floors and on the bases of pots.

Wool was universally used in the Near East and Europe except in Egypt, where flax grew from which linen was made. In Europe, a fine linen-like cloth was made from nettle stems. Vegetable dyes were used – madder for red, weld for yellow, woad for blue.

The first evidence of looms comes from representations on a seal from Iraq and a dish from Egypt, just after 3500 BC. There are horizontal looms like those still used in many parts of the world, the weaver squatting on the ground. The upright loom came into use in the Near East during the second millenium BC. This took up less space, and could have been used indoors. Upright looms are shown on a pot from Sopron in Hungary, dating from the 6th century BC, and on Greek pots of the 5th century BC.

A saddle quern with rubbing stone, about 52 centimetres long.

PREPARING AND EATING CORN

The grains of wild cereals are not suitable for making into bread. Probably the early husbanders made it edible by parching – heating it in an oven. This removed the chaff and made the grain more edible. Parched grain could have been ground in a quern and made into a gruel; such querns are common in Near Eastern sites. Gruel left to stand for a time will often ferment, forming a kind of beer which may well have been drunk from early husbanding times on. Bread was probably not made until wheat was cultivated and the grains contained enough gluten to hold the dough together and retain gas.

	AFRICA & EUROPE	WESTERN ASIA	FAR EAST	AMERICAS
3000	First farming, Northern Europe			No villages until 1500 BC
4000	First farming, Central Europe First farming villages, Egypt			
5000	Lepenski Vir: 5400 BC	First settlements in Sumer; first irrigation	China: Yang-shao culture, first villages	Cultivated maize, Mexico First cultivated plants, Peru
6000	First farming, SE Europe	Permanent villages; domesticated animals and plants		First cultivated plants, Mexico
7000	Domesticated cattle, Greece: 6500 BC	Seasonal villages; 'husbanding'	Thailand: Spirit Cave, domesticated plant remains?	First domesticated dog—Idaho, USA: 8400 BC
8000		First domesticated sheep?—Shanidar		

able materials as skins and gourds would also have been used. Jarmo in Iraq was another village of this time; it was only a quarter of the size and housed some 150 people in houses built of clay and gravel with mud-covered reed roofs.

As well as meat and cereals, legumes formed an important part of the diet at these villages, but they became less important as cereal production increased. Food gathering continued, and pistachio nuts, fruits, snails, and waterfowl were eaten. Sheep and goats were domesticated, the goat being more important. At Jarmo and at Çayönü in Anatolia pig bones from successive levels are said to show stages towards domestication. There were domestic dogs at both these sites, which would have been useful for herding.

At the village of Ganj Dareh in Iran, 8 metres (26 feet) of deposits accumulated in this phase, showing a long occupation. Level D, with a radiocarbon date of 7018±

100 BC, was destroyed by fire, which baked and preserved the walls of rectangular houses constructed of long mudbricks. An upper storey was supported on wooden beams. The fire also preserved the oldest known pottery, small vessels and large storage jars. Clay may have been shaped into pots before this time, but only sun dried or lightly fired; and unless pottery is fired in a kiln at a temperature above 400°C it crumbles without trace.

Down to the valleys

In Iraq a significant advance came in the sixth millenium BC when people moved from the hill country where cereals grew wild into the river valleys of the Tigris and Euphrates. (It was only in the fifth millenium that agriculture spread into Egypt and the rich Nile valley.) Such a move meant adapting cereals to a lower altitude and richer soils. Important sites are Hassuna and Yarim Tepe in north Iraq, and

A fragment of carbonized cloth, found in one of the shrines at Çatal Hüyük.

A house excavated at the neolithic site of Abu Hureyra in Syria. Built of mudbrick, it dates from the 7th millenium BC.

RICE

Rice (*Oryza*) is the staple food of over half the world's population, although it contains even less protein than wheat and maize. There are about 25 species; the probable ancestor of *Oryza sativa,* the most widely grown today, is the wild *Oryza perennis* found in humid tropics. Another domesticated species, *Oryza glaberrima,* has been grown in West Africa since 1500 BC. Although there have been few finds of rice from ancient sites, it is probable that wild rice was collected and husbanded by early man in the same way as other grasses, mutants being selected for larger grains and for less brittle stalks. The oldest archaeological evidence of rice comes from Thailand and dates from about 5000 BC. Rice was domesticated in centres from the Ganges plains across Burma to South-east Asia.

Samarra to the south near Baghdad. Many of the sites of this phase are south of the limit where rain-fed agriculture would have been possible, which implies the earliest development of irrigation – perhaps beginning with the damming of seasonal floods. Villages were defended with walls and towers. In the extreme south of Iraq, the al 'Ubaid culture arose with its beautiful painted pottery; its early stages have been revealed by excavation in the lowest levels of the mound or tell at Eridu. Important evidence has come to light here of a continuity of religious tradition, for the plan of much later Sumerian temples can be traced back to mudbrick buildings of this very early stage.

Çatal Hüyük

In Anatolia the excavations by James Mellaart at Çatal Hüyük revealed a similar stage beginning about 6500 BC. Farmers kept sheep and goats, and hunted aurochs, wild pig, and red deer as well as other deer and leopards. More than 14 different food plants were cultivated, the most important being emmer and einkorn wheat, naked six-row barley, and peas. Lentils, vetches, and shepherd's purse were also grown, and almonds, acorns, pistachios, apples, junipers, and hackberries were gathered. In the small area of the mound that was excavated were two types of buildings, houses and shrines, distinguished mainly by the decoration that had survived on the walls. Houses were decorated by geometric patterns and panels of red paint. But in shrines vivid scenes with animals and human figures were accompanied by animal skulls fastened to the walls, and plastered to simulate the living heads. A pair of leopards on the walls of one shrine were covered with 40 layers of paint; the shrines and houses were apparently replastered and repainted every year. The dead were buried inside the houses and shrines, in platforms on top of which the inhabitants probably slept.

The accidental burning of some levels preserved traces of wooden beams and

The largest neolithic site in the Near East is Çatal Hüyük, on the Konya Plain of southern Anatolia. It was occupied by early farmers for over 7000 years, and the oldest levels found so far are dated to 6385 BC. Çatal was a thriving town, only a small part of which has so far been excavated. Houses and shrines were built of mudbricks and mortar, round a framework of large timbers, all shaped with stone tools. There were no streets, but courtyards between the houses. Access was across the flat roofs and down ladders. Disastrous fires swept the town at intervals, baking the mudbrick walls so hard that they could not be levelled. The rooms and courtyards were filled up with rubbish (from which comes evidence of pottery, food, and tools) and new houses built on top. Buried walls have been preserved to a height of some 3 metres and on them have been found many wall-paintings, modelled female figures, and animal heads modelled over skulls. Paintings included vultures and headless men and patterns similar to those still used in the region; the walls were replastered every year, and new paintings superimposed on the old. Leopards were modelled on the walls and little figurines have been found of women with leopards; they may have been kept as sacred pets. The dead were placed on exposure platforms, and their bones were later collected and buried inside the low platforms on which families sat and slept. Wooden bowls, necklaces, and other possessions were buried with them, and inside one skull was found a length of finely woven wool or mohair cloth.

fired the mudbrick walls so hard that they could not be pulled down. Later buildings had to be constructed on top of existing walls, the rooms being filled with rubbish that has proved invaluable for reconstructing everyday life. Fires also baked and preserved cloth, baskets, and wooden vessels buried in platforms with the dead.

Early Europeans

In Greece, Crete, and the Balkans, the climate was similar to that of the Near East and farming developed there almost as soon. As in the Near East, the earliest remains represent an *aceramic* (without pottery) stage. The house walls and rubbish of people living on the same site for

many generations accumulated in great mounds, like the tells of the Near East.

Much information about the first European farmers has come from neolithic levels at Knossos (beneath the later Palace of Minos), at Argissa in Thessaly, and at Karanovo in Bulgaria – a mound over 12 metres (40 feet) high which was occupied between 5000 and 2000 BC. These farmers of south-east Europe used bone spatulae (possibly for lifting flour) identical with those found in Anatolia, and flint sickles for reaping. They made small clay figurines of women – rarely of men – and their pottery, in use after 6000 BC, was well made. During the third millenium farming spread into Romania and south Russia. In the Cucuteni and Tripolye cultures long communal family houses of wattle and clay or of beaten earth were grouped in villages of 30 to 50 dwellings. Pottery vessels were painted with elaborate spirals and animal figures. Horses were domesticated.

During the fifth millenium BC farming spread north of the Danube into colder and wetter areas where mud-walled, flat-roofed houses had to be replaced by log-cabins with gabled roofs. These farmers made the first pottery in central Europe; it is decorated with spirals tooled on the damp clay and has led to the term Linear Pottery Culture for its makers. These farmers moved rapidly across Europe, cultivating the lightly wooded loess soils. With polished stone axes they cut down and burned the tree cover; this made the soil extremely fertile for a few harvests, after which the clearings had to be abandoned, since they did not know how to manure the soil to preserve its fertility. Domestic animals grazing in clearings prevented the trees from regenerating. This 'slash and burn' agriculture contributed to the gradual deforestation of much of Europe.

The first farmers in northern Europe – in north Germany and Denmark – were the Funnel Beaker Cultures from around 3000 BC. Named after their characteristic pottery, they represent the adoption of farming by native food gatherers. The beginning of farming, here as in the British Isles, is marked by a drop in the pollen of elm trees as seen in samples from peat bogs.

Mediterranean spread

During the fifth millenium BC farming also spread westwards through the Mediterranean, from Sicily to Spain. Small groups of people settled along the shores, living in caves and decorating their round-based pots with impressions of cardium shells. They have been named the Cardial Cultures. From these beginnings arose the temple-building farmers of Malta in the third millenium BC. In France, farmers moved northwards from the shores of the Mediterranean, making dark, round-based bowls, similar to those found in Britain during the fourth and third millenia BC.

Exploiting Metals

Man has used metals since far into prehistoric times, long before the invention of writing. So we know about the earliest metallurgists in many areas only through archaeological excavation. This has brought to light not only metal artefacts – tools, weapons, vessels, and ornaments – but also parts of the apparatus of manufacture: ingots, moulds, crucibles, furnaces, and slag. The examination of these by metallurgists has thrown a good deal of light on the early stages of metal working in different parts of the world.

While many metal artefacts are isolated finds, others have been excavated from the remains of houses and domestic refuse, or from graves. On many sites, these remains form layers with the older material at the bottom and the more recent above. For example in Denmark, during the last 2000 years BC, metal artefacts were thrown into lakes as offerings to gods. Peat has since formed in these lakes, preserving layers of metal objects in the order in which they were thrown.

From such stratigraphical evidence from many sites, it has long been clear that the first metals used in the Near East and Europe were copper, gold, and silver. Bronze, an alloy of nine parts of copper to one of tin, is found in higher levels, as smiths learned to get round the difficulties of casting pure copper. Iron objects are everywhere separated from the first bronze artefacts by levels representing a thousand years or more – an interval which reflects the difficulty of smelting iron from its ores.

On the basis of this stratigraphy, Danish scholars of the early 19th century classified primitive man's cultural development in three stages: a universal 'Stone Age', followed in certain parts of the world by a 'Bronze Age' and an 'Early Iron Age'. Although these terms are now being superseded by periods dated in years, the sequence of such stages, with a copper-using stage preceding that of bronze, is valid for all the areas where metalworking developed.

Right: Azurite, a copper ore whose brilliant colours would have made it attractive to primitive men.

Above: An iron mould for casting iron sickle blades, from Hopei in China; it dates from the 5th to 4th century BC. The Chinese evolved a method of casting iron soon after they discovered iron-working; but in Europe the metal could only be shaped by hammering until the blowing engine was invented in the 12th century AD.

The origins of metalworking

The oldest centres of metalworking are in the east Mediterranean area. They are close to deposits of native metals, which could be used without previous smelting from parent ores. The brightness of copper, of silver (only rarely found in a native state), and of gold would catch the eye of people searching for stone for tools.

Native metals can in fact be shaped by striking them with a hammerstone, as flint and other stones had been shaped from time immemorial. Iron from meteorites – which can be distinguished from terrestrial iron by its high nickel content – was occasionally hammered into trinkets like those dating from the fourth millenium BC in Egypt long before an 'Iron Age' stage of culture was achieved. Even

	EGYPT	EUROPE	WESTERN ASIA	FAR EAST
1000	Middle Kingdom; bronze in common use	Hallstatt culture—first iron, N Europe: 700 BC Bronze and gold in common use	Hittites—first iron: 1500 BC Bronze, gold, and silver in common use	First wrought and cast iron, China: 500 BC First metallurgy in China—bronze and copper: 1500 BC
2000	Old Kingdom First tin bronze: 2680 BC	First copper, bronze, and gold in C and W Europe	Alaca Hüyük tombs: 2400–2200 BC	First bronze in SE Asia
3000	Predynastic stage; use of native copper, silver, gold	First bronze in Aegean Copper shaft-hole axes in Hungary First smelted copper in SE Europe	First cast copper, Iran and Sumer	
4000		First cold-hammered copper in SE Europe	Pottery kilns, Iraq	
6000			First cold-hammered copper, Anatolia	

in the early 19th century AD Eskimos in Greenland, who had no knowledge of smelting, were battering small fragments of iron from a meteorite and setting them in hafts of walrus ivory to make knives. This was using iron simply as mesolithic hunters in Europe 10,000 years earlier had used flint microliths.

Gold, from veins in quartz rock or from stream beds where it had been carried through erosion, was being hammered into small ornaments before 4000 BC in the Near East. But native copper was being worked into small artefacts, pins, beads, and awls long before this date in Anatolia and Iran, where the main sources of metals for the Near East are found. The oldest metal artefacts in the world that we know of are those excavated in 1968 at Çayönü Tepesi in south-east Anatolia – an ancient village only 20 kilometres (12½ miles) from one of the major mining centres of

Copper was being extracted from deep mines in the eastern Austrian alps from about 1200 BC. Ancient workings near Salzburg have been extensively excavated and were found to extend 160 metres (175 yards) into the mountain. Since there were then no explosives, fire was used to make the tunnels. A fire was lit against the rock surface, and, when this became hot, cold water was thrown on to the rock to make it split. As the tunnel deepened, the roof was supported with big wooden pitprops. After a time, the tunnel was high enough for a solid wooden ceiling to be constructed over the miners; space was left for the waste quarried rock to be placed between the wooden ceiling and the roof of the working. Miners had different tasks. Some worked at night controlling the fires, and fetching wood and water from mountain streams or from sumps in the workings. After the fires had been put out and the smoke had cleared, the day shift of miners hacked out the ore from the working face with stout bronze picks; others crushed it with huge stone hammers, so that the shiny ore could be picked out from the rock. The ore was carried outside to be crushed and washed out from the last of the rock. The concentrated ore was taken farther down the hillside and smelted in a clearing, where traces of furnaces have been found.

The cold damp atmosphere of the mines has preserved many wooden objects such as notched ladders, troughs, and tool handles, so that it is possible to reconstruct in detail the life of the miners. Even bundles of twigs were found of the kind that were burned to provide light in the dark tunnels of the mines.

modern Anatolia. These hammered metal objects include three pins, a small tablet, beads, and the point of a reamer (a tool used to enlarge or shape a hole), and they have been dated to just before 7000 BC.

The makers of such small, cold-hammered artefacts cannot strictly be described as 'metal-using'. This term is used only for a society that has learned how to smelt metal from its ore and to shape objects by casting in moulds. Such activities use the special properties of metals through the application of heat – a very much more advanced stage than the use of metal as a superior kind of stone.

True metalworking began only after it was realized that heat could be used for two simple purposes; melting and annealing. Melting in a container or crucible would enable smaller pieces of metal to join into a larger and more useful amount. Some metals – including copper – become brittle when hammered for any length of time. Breaking is only avoided by annealing – heating the metal at intervals.

Extracting metal from ores

The first metals to be used in the Near East, from the seventh millenium onwards, were gold, copper, and, to a smaller degree, lead and silver. Gold was used for jewellery, and for splendid vessels such as those in the third millenium BC Shaft Graves of Ur, but it was too soft to be used for tools and weapons. Copper was used for these utilitarian purposes, although bone and stone continued in use side by side with metal until the adoption of iron.

Most copper is found in the form of ores, from which it must be extracted by smelting in a furnace with a temperature of at least 700°–800°C. Although the temperature of a campfire is normally less than that required to smelt copper, reaching 600°–700°C, a geologist in Zaire once found that copper had been accidentally produced in his campfire; it had formed in beads on the surface of lumps of ore which he had used to build a fireplace. A similar happening could have shown an observant person in antiquity the immediate source of the pinkish-orange metal.

The furnace required for smelting was probably derived from the potter's kiln. It is possible to work out from the fabric of a pot the temperature at which it was fired. In the Near East and in the Balkans pottery has been found that must have been fired in kilns at a temperature high enough to smelt copper. A simple form of

Above: A bronze lur (trumpet) found in a peat bog in Denmark. It was cast by the lost-wax process in segments which were then joined. It is some 1·8 metres long, and its shape is derived from the horns of wild cattle.

bellows could have been used to raise the temperature.

The temperature of a furnace could also have been increased by directing the wind into it. For iron smelting in Germany (and possibly earlier for bronze) furnaces were dug into the side of a hill with the opening facing the direction of the prevailing wind. The draught was forced through the fire and up a vertical chimney which in the Jura type of furnace of the first millenium BC was some 3 metres (10 feet) in height.

Once metals had been smelted, smiths discovered how to cast them in moulds. The oldest known moulds in the Near East and Europe were one-piece open moulds, the shape of the desired object being cut into one side of a block of stone. Molten metal was poured in, and the mould was covered with a wooden lid to keep out as much air as possible and prevent bubbles forming in the casting. One disadvantage of the open mould was that one side of the artefact was inevitably flat, and could only be shaped by forging after casting. To make an artefact shaped on both sides, a two-piece mould was developed, made of stone, clay, or, later, of metal.

To make a hollow object, such as a socketed spearhead, a core (often of clay) had to be suspended inside the mould, which was pegged together round it.

Very elaborately decorated objects could be cast of copper, bronze, silver, or gold by the 'lost-wax' or *cire-perdue* process. A beeswax model of the exact form of the object was coated with fine clay, which ran into all the finely tooled details of the modelled surface. When this had dried, a coat of thick clay was applied. A hole was left in the clay at the top or bottom of the mould, and when it was heated the wax would melt and run out of the hole, leaving a cavity inside the mould. Molten metal was then poured in, and replaced exactly the lost wax original; but the mould would have to be broken open and destroyed to reveal the casting.

This sophisticated technique was used

Above: The lintel of the Sumerian temple of Nin-Khursag at Al Ubaid, Iraq; it dates from the early fourth millenium and is made of copper sheet, wrought by hammering and annealing. The lion-headed eagle Im-dugud links two stags. It is 2·4 metres long and just over 1 metre high.

EARLY TECHNIQUES

In the Balkans the first stages of metal-working — cold-hammering native copper into simple pins, fish-hooks, and bracelets — are now dated to before 4000 BC. It seems that during the fourth millenium BC two local forms of axe were made: the axe-adze (a kind of pickaxe) and the shaft-hole axe, both of which have been found in considerable numbers. They were made by people living in established villages of 20 to 30 families, living by efficient farming in fertile river valleys. Their pottery included vessels decorated with graphite, which must have been fired in sophisticated kilns reaching a temperature of 1050°C, in a deliberately controlled reducing atmosphere. These conditions are identical with those required for smelting and casting copper, so that, although no furnaces or workshops of fourth-millenium smiths have been found in the Balkans, copper axes could have been made by casting and not by cold-hammering from native copper. This has now been proved by metallurgical examination.

Skilled as Balkan smiths were, they did not discover how to alloy copper with tin to make bronze until the third millenium BC. They could only increase the hardness of the copper by hammering the edges after casting.

A great deal of information about ancient metalworking has been gained from metallurgical examination of metal artefacts — firstly about the source of the metal employed, and secondly about the methods used by smiths. These included the use of alloys and the actual techniques of manufacture, such as cold-hammering, casting, forging, and hardening of the finished object by cold-working. Tiny amounts of metal are removed from the artefacts for analysis; the minute holes made by this are filled with hard wax and painted over so that the object's appearance is not spoiled.

The oldest artefacts in the Near East and Europe are such basic, simple forms as beads, pins, and awls, and — at a more advanced stage — flat axes. Archaeologists need to know whether these are made of native copper or of copper smelted from ores, a later stage of metal-using development.

It is not always easy to distinguish between native and smelted copper, although a basic difference is that native copper is normally pure, while copper smelted from ores can contain trace-elements of other minerals, including arsenic, antimony, silver, nickel, and even tin. These traces vary from one deposit to another, and when more is known about the composition of ores from individual deposits it will be possible to work out the distribution patterns of ores and artefacts through trade.

These axe-adzes, made in the Balkans during the fourth millenium BC, are the oldest-known smelted copper objects in Europe.

in the third millenium BC in Sumer to make such complex models as the copper chariot group found at Tell Agrab, the oldest known example of lost-wax casting. It was discovered independently in China and in South and Central America.

Bronze

In all the ancient metal centres, the disadvantages of casting in pure copper led smiths to experiment in alloying it with other minerals – experiments that must have been stimulated by finding that different results were obtained by the use of copper from different lodes, some of which were naturally alloyed. By 3000 BC, the time of the first cities in the Near East, a standard formula of nine parts of copper to one part of tin had been worked out, making the alloy known as tin bronze. The addition of even this small amount of tin made the metal easier to cast and less likely to be damaged by air bubbles, and the finished product was a harder metal; but the tin was hard to come by. Compared with copper, tin is rarely found, and in Europe the deposits of Iberia, Brittany, Cornwall, and Central Europe brought rich and exotic objects to these areas in exchange for the precious commodity. The far-flung trade in tin must have made bronze artefacts expensive, which explains why the much more common metal, iron, replaced bronze for tools and weapons after the problems of extracting it from its ore were overcome.

Bronze was discovered independently in the Near East, in China, and in the New World. One strange aspect of Chinese metallurgy is its sudden emergence about 1500 BC, with no gradual beginnings like the hammering of native metal in the Near East. This led some archaeologists to infer that the knowledge of metallurgy must have reached China from the Near East, along the route that formed the much later 'Silk Road'. But in some ways, Chinese technology is completely different from that of the Near East. The oldest known moulds, for example, are not single-piece, but are highly complex moulds of up to 30 pieces. Each was elaborately shaped, so that the casting required very little finishing. Bronze containing a high proportion of lead was used; the low melting point of the lead allowed the metal to run into all parts of the elaborate

moulds before it solidified. Lead bronze was used in a similar way in Britain in the last millenium BC – but in simpler moulds.

In China as in the West, the smelting furnace was probably derived from the potter's kiln, for Chinese pottery of 1500 BC must have been fired at a temperature of 1200°C. The shapes of bronze vessels are very close to contemporary pottery.

Iron

Iron was first used for tools by the Hittites of Anatolia about 1500 BC. It is much easier to obtain than copper or tin, since it occurs in many more places and can be picked up on the surface or obtained from opencast working. The long interval that elapsed in all areas between the discovery of copper metallurgy and the use of iron can be explained by the difficulty of extracting iron from its ore, which contains sulphur. This has to be expelled by roasting the ore in a slow fire with charcoal. Another disadvantage of iron is that it could not be cast in the furnace used for melting copper or bronze. Iron does not become molten under a temperature of 1535°C, and the heat of the furnaces of the Near East and Europe was only 900°–1200°C. At these temperatures, copper would melt out of its ore and trickle down to the bottom of the furnace to form a

Above: A copper stag, 20 cm high, probably cast by the cire-perdue or lost-wax method. This was found on the bier of a dead king at Alaça Hüyük, central Anatolia, and dates from 2400 to 2200 BC.

Left: One of a hoard of five gold torcs found at Ipswich on the east coast of England. It was made from a faceted gold bar which was bent back on itself and joined to form an elongated loop; the loop was then twisted and bent into a ring shape. The ends had relief ornament cast on to them. The torcs were made by early Celtic smiths and worn round the neck.

A bronze ritual wine vessel, or yu, made in China during the 11th or early 10th century BC. The elaborate decoration includes stylized birds and dragons. Such complicated shapes were made possible in moulded work by the addition of lead to the bronze; this helped the liquid metal to flow into all parts of the mould. The handle swivels to allow the lid to be removed.

107

Above: A bronze cult-wagon found in a Hallstatt culture barrow at Strettweg in Austria; it dates from possibly the 7th century BC. The goddess in the centre (22·5 cm high) carries a vessel on her head and is surrounded by mounted warriors, figures on foot, and two stags.

Below: One of a pair of wrought-iron firedogs from Capel Garmon in Wales, dating from the 1st century BC. Depicting an ox head, it is 24 cm high.

bun-shaped ingot. Iron would not melt, but the ore would change into a solid but spongy mass of iron mixed with slag, known as a 'bloom'.

The bloom had to be taken from the furnace and hammered red hot, on an anvil. The iron in it was gradually hardened through the addition of carbon from the charcoal furnace, while in the course of repeated heating and forging the slag flew off in the form of sparks. The smith knew by experience when all the slag had been expelled, and then forged the iron into tools, weapons, or objects for the home such as firedogs.

Only such wrought-iron artefacts could be produced in the Near East and Europe until the 14th century AD, when the blowing engine was invented and the temperature of furnaces could be increased to make iron molten. But in China, although the use of iron in any form did not begin at all until the 5th century BC, the earliest iron objects include not only wrought but also cast tools, and cast-iron moulds for making bronze artefacts. Furnaces capable of extremely high temperatures must have been constructed there nearly 2000 years before iron could be cast in Europe.

Metalworking and society

The discovery of metals, particularly iron, brought increasingly mass-produced tools and weapons into everyday life. Made by specialists, tools became available for many different crafts and activities, making life easier in many ways for those who practised them. Hammers and metal anvils and tools for ornamenting metal aided the smith; the carpenter acquired hammers, chisels, and gouges as well as saws. The saw made planks quicker to obtain, replacing laborious work with the adze; it was only after the development of such metal tools that the light spoked wheel for carts and chariots could replace the heavy, solid wheel. The wheelwright could also increase the life of a wheel by encasing it in an iron tyre. Farmers were helped by bronze – later iron – sickles and iron ploughshares.

Individual hunters and warriors had always provided themselves with lethal spearheads and arrowheads of stone, but metal armaments could be mass produced from moulds or – in the case of iron – by forging. A broken stone weapon had to be discarded, but a metal one could be melted down to make a new one. The discovery of iron was particularly important in equipping large armies cheaply with weapons and with slave chains for captives.

In the home, food could be boiled or stewed in metal cauldrons suspended directly over the fire – much simpler than the older method of heating a stew by dropping hot stones from the fire into the pot. After a feast or at religious ceremonies bronze wind-instruments sounded; the shape of 8th-century Irish instruments suggests that these replaced earlier ones made from the horns of cattle.

Gold and silver were from the beginning prized for their beauty. Necklets, armlets, pins, brooches, and so on were not only ornaments, but a means of accumulating and demonstrating wealth and rank. Metal also provided a currency that did not die – a disadvantage of such primitive forms of barter as cattle and slaves.

Metalworking affected not only man's everyday life but also his environment. Pre-metal neolithic farmers had begun the process of deforestation, making clearings in the primaeval forests for their fields, and this continued on a much larger and faster scale as metal axes spread into different areas of the world. And the trees were now felled not only to clear room for fields and to provide fuel for cooking fires and pottery kilns, but also to feed the smelting furnaces. Man's heedless devastation of the world around him had begun.

Advanced Societies

By 5000 BC there were large, prosperous 'towns' and villages in many parts of the Near East. Jericho in Palestine, for instance, covered 4 hectares (10 acres) and must have had a population of several thousands. An efficient economy, based on food production, had been achieved at sites from Palestine to Anatolia and east to Iran; but the next stage, that of city life – the literal meaning of 'civilization' – was not reached for over 1500 years. The first true cities, such as Ur and Eridu in Iraq, arose not in areas where the revolutionary discoveries of farming had been made, and where the oldest villages and towns grew up, but in the fertile river valleys of the Tigris and Euphrates, the Nile, and – later – of the Indus.

In the highly organized life of a prehistoric city, the activities of the inhabitants were far removed from basic food production. The food was grown by farmers outside the city walls, while the city dwellers organized its distribution, and engaged in trade; many were craftsmen, whose work was sold by shopkeepers. Others were architects who designed public and religious buildings. Rulers governed the city, with a staff of clerks, scribes, and administrators. The worship of the gods was directed by priests and their acolytes.

None of these people produced their own food. Instead they had something to offer the community in return for the food they consumed; they could only be supported if the farmers outside the city could produce a surplus of food far beyond their own needs. Such a surplus simply could not be achieved by the inhabitants of the first neolithic villages and towns; it was only in the incredibly fertile river valleys that huge quantities of corn could be grown to feed an increasing number of non-farmers.

In Egypt and Sumer, stages towards the development of cities can be traced. First there were the villages of neolithic farmers. Then came a *predynastic* stage, when a surplus was becoming available to feed non-producers and writing, trade, and industries such as potting and metallurgy

This steatite statue, possibly of a god, priest, or king, comes from Mohenjo-Daro and dates from about the end of the third millenium BC. The eyes were originally inlaid with shell and the tre-foil decoration filled with red paste. The trefoil was a religious symbol connected with the stars, and is also found in Sumer. The statue is 17 cm high.

were beginning. Lastly emerged the full city life, the *dynastic* stage, when in about 3000 BC dynasties of powerful rulers can be identified.

One factor unifying the population in river-valley settlements was the over-riding need for social co-operation in irrigating the fields. Individual ditch-digging was useless. An overall plan had to be devised by a strong leader, who became the chief or king. To bring success to the irrigation the gods had to be propitiated with sacrifices and rituals. Thus a priesthood grew up, and the construction of elaborate temples led to the development of architecture and mathematics. So that the annual floods could be prepared for, a calendar was worked out – in Egypt based on the rising of the star Sothis (Sirius); in Sumer, on the movements of the planet Venus.

Cities of Iraq

The first cities in Iraq grew up in the Euphrates valley. Their development has been traced through the excavation of *tells*

– mounds formed of superimposed ruins of mudbrick buildings. In the lowest levels of such sites as Ur and Erech traces of neolithic villages have come to light. This primitive stage has been named the Ubaid culture; it lasted from about 4000 to 3500 BC. During the following predynastic stage from about 3500 to 2900 BC, the villages can be seen to develop into much more complex settlements, with .larger and more elaborate temples. This was under the stimulus of the people of the Uruk culture, who may have come from Iran, and mixed with the older population. A number of centres grew up on sites that later became famous cities, such as Nippur. The inhabitants had to work together under a strong leader. Sowing had to be undertaken in the spring, and dykes dug for defence against the early summer flooding of the Tigris and Euphrates which would damage the growing crops, as well as for irrigation. The dread of floods and the havoc they caused coloured early Sumerian literature, in deluge legends that inspired the biblical Flood story.

The development of the plough improved the yield of corn, ensuring that seed was sown more evenly in rows which could be weeded. In the south of the Euphrates valley the diet of corn and meat from the farms could be supplemented by fishing and fowling in the marshes of the river deltas.

The cities of the Euphrates valley had no natural resources at all except the rich and fertile mud. This was shaped into

Above: A clay tablet in its protecting envelope; the cuneiform writing points out that there has been no reply to two previous letters. It dates from 1700 BC.

A tablet found at Kish, Iraq, bearing pictographs. It dates from about 3500 BC, and is thus one of the oldest-known written documents.

CUNEIFORM

Writing began in Mesopotamia in the predynastic Uruk period (from about 3500 BC) in the form of symbols cut into seals used to stamp an individual's possessions. With the growth of trade, records of transactions were increasingly needed and writing developed further; at Kish and other sites clay tablets have been found bearing vertical columns of signs drawn with a pointed tool. These are pictograms, stylized drawings of the object concerned; some of them, including the signs for star, water, and earth, have been recognized on older painted pottery.

A pictogram system of writing requires an enormous number of different signs. The Sumerians progressed to the use of ideograms (signs for an idea associated with a particular object) and phonograms (signs representing sounds). These reduced the number of signs needed, but could lead to ambiguities. The sign for 'water', for example, came to represent the sound *a*, the Sumerian word for water. But the sound *a* could also mean 'in' – so the same sign had two meanings. A new class of signs, called determinatives, was invented to show the class of word spelled by the phonograms.

In time, the signs became simplified so that they looked less like drawings of real objects but were quicker to write. Further changes occurred as scribes began to use a pen of triangular section cut from a reed. The wedge-shaped marks left by this method of writing give the name of the script as a whole – 'cuneiform', from the Greek word for wedge. Cuneiform developed in the third millenium BC, and continued in use until the first millenium BC for many different languages, including Sumerian, Akkadian, and Hittite. Libraries of cuneiform tablets have been found at many sites, and inscriptions have also been found carved on stone.

Cuneiform was deciphered independently by two scholars; by the German Georg Friedrich Grotefend in 1802, and the Englishman Henry Rawlinson, between 1844 and 1847. They began by studying monumental inscriptions in Old Persian, and identifying the names of known kings.

Above left and above: The so-called 'Standard of Ur' – possibly an ornament for the sounding box of a lyre – is about 46 cm long. Found in the Royal Cemetery, it was made about 2700 BC. Figures of shell and red limestone are set in a lapis lazuli background, fixed to the wooden box with bitumen. On the first side the upper row shows figures feasting to the accompaniment of a lyre and singer (right); in the middle and lower rows men are driving animals and carrying heavy loads by means of a head-band. On the side shown above are scenes of war, with heavy four-wheeled chariots whose solid wheels were made of two separate pieces of wood fastened together. They are drawn by wild asses or onagers. In the upper row, the central figure of the king has left his chariot to inspect prisoners who are being led before him. In the middle row on the left is a group of soldiers with helmets and cloaks. On the right the vanquished are being made captive. The bottom row shows a procession of chariots, each charioteer accompanied by a soldier, driving over the bodies of the slain.

building bricks which were dried in the sun, not fired, so the surface weathered away. The surfaces of temple walls were protected by cones of fired clay, some with heads of coloured stone, which were hammered into the surface in colourful mosaic patterns. The lack of long stone blocks for lintels led to the development of the arch; true arches and barrel vaults were used in the construction of early dynastic burials at Ur about 2600 BC. Among the imports of early cities were raw materials from areas as distant as Anatolia and India: wood and

stone for building, semi-precious stones such as lapis and carnelian, gold and other metals. These transactions were recorded on clay tablets, using first pictograms and later cuneiform writing. Documents show the structure of city society in early dynastic times. The king ruled from an elaborate palace; the city god owned one-eighth of the city's lands, which was administered by the priests. Other land was owned by great families. There were also landless freemen, and slaves who were prisoners captured in the many wars between city

Left: The crushed skull and gold head-dress of one of the many court women who were buried with the king in the death-pit at Ur. Sir Leonard Woolley preserved these and other fragile finds by embedding them in wax, so that they could be taken away from the site for reconstruction. Far left: A similar reconstructed head-dress, made of gold foil and lapis lazuli. Above: A he-goat resting on a tree, one of a pair of figures found in the death-pit. It is made of wood, mostly covered in gold sheet. The lapis horns are rivetted to the head, and the locks of wool, also lapis, stuck to the body with bitumen. Above left: A Mesopotamian stone cylinder-seal of about 2300 BC. Its impression shows two bulls, each being stabbed with a dagger by a bull-man.

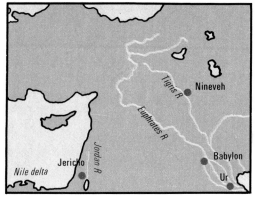

The area around Mesopotamia is sometimes called the Fertile Crescent, because the mild Mediterranean climate and rich soil around the rivers made heavy crops easy to grow. This ensured a surplus every year, and non-producers could be supported. This made possible a higher degree of civilization than ever before.

	EGYPT		EUROPE	ASIA
1000			Etruscan civilization	Phoenician civilization Persian civilization
		c 1575 New Kingdom	Height of Mycenaen empire c 1700 Height of Minoan civilization	1500 End of Indus civilization
2000		Middle Kingdom	Tell-villages in C Europe	
		Old Kingdom		c 2500 Indus civilization Cities in Near East
3000		c 3200 1st Dynasty		Predynastic stage, Iraq Uruk culture, Sumer
4000		Predynastic stage		Ubaid culture, Sumer

states. A distinctive Mesopotamian civilization continued until the 5th century BC, when the Persians conquered the area.

In the valley of the Nile, civilization arose in stages parallel with developments in Iraq. The growth of this Egyptian civilization is described on page 135.

Cities of the Indus

In 1921 traces were found in the Indus valley in the north-west of the Indian peninsula of another ancient centre where civilization arose about 2500 BC. The origins of this are mysterious; knowledge of farming may have come from the highlands of Iran and Iraq, but in addition to animals kept in Sumer, the Indian farmers kept humped cattle, domesticated the buffalo, and tamed the elephant, as well as being the first farmers in the world to grow cotton.

Nothing is known of the organization of life in the Indus cities, since no documents have survived, and the strange pictograms on pots and seals cannot be deciphered. But the city-dwellers were fed here as elsewhere by farmers in the surrounding countryside, who brought their grain in two-wheeled carts to huge mudbrick granaries on the citadel mounds at Harappa and Mohenjo-Daro, the two largest cities. Other buildings on the Mohenjo-Daro citadel included a great bath or storage tank; a group of cells with private baths, perhaps for priests; and assembly halls. Both the major cities were about five kilometres (three miles) in circumference. Houses were laid out in

blocks – evidence of careful town-planning. Main roads were five metres (16 feet) wide, and were unpaved; brick-built drains were evidently cleaned out at intervals as access for inspection and cleaning was provided by means of manholes. There were public wells in addition to some in private houses. Houses had no

A seal, made in the style of the Indus valley but found in Mesopotamia, shows that trade contacts between the two civilizations must have existed.

Above: The helmet of Mes-kalam-du, found on his skull in the Royal Cemetery at Ur. Its shape follows the hair-style of its wearer. There were traces inside of a quilted cap. The helmet was hammered from a single sheet of gold.

Below: The Great Bath at Mohenjo-Daro after excavation. The brick steps at each end originally had wooden treads set in bitumen. The bath itself, measuring 12 × 7 metres, was waterproofed with asphalt.

A *terracotta model – perhaps a toy – of a cart drawn by two oxen, from the Indus valley. The heavy, solid wheels represent the three-piece wheel, the earliest type of wheel known, which probably developed in Mesopotamia. The model dates from about 2000 BC.*

windows on the outside walls, but consisted of rooms built round a courtyard. The door opened on to a side alley rather than a main road, along which were shops.

Harappa and Mohenjo-Daro are the largest Indus cities known, but smaller towns, without citadels, have also been found. No royal graves have been found but at Harappa one body was buried in a coffin of scented rosewood, and cemeteries have been found where people were buried with up to 20 pots in one grave.

Little stone or bronze sculpture has survived, but many terracotta statuettes have been found, some possibly toys. Others – bulls, hens, and female figurines – may have been votive. The most striking art is portrayed on the stone stamp seals, on which designs were deeply incised with chisels and drills.

The Indus valley people traded with the centres of civilization in the Near East; beads from Egypt and Iraq have been found at Mohenjo-Daro and Harappa, and objects of Indus valley manufacture have

Right: Among the very few bronze sculptures found in the Indus valley is this life-like dancing girl, found in a house at Mohenjo-Daro. Her hair is elaborately dressed and she wears a necklace and many bracelets.

The Granary at Mohenjo-Daro was built of solid timber on a huge brick base. The cutaway reconstruction below shows how the base was built in 27 sections, allowing air to circulate through the gaps between.

been found in Sumerian cities. The use of mudbrick for building, the development of public irrigation and flood-control systems, the smelting of copper and later of bronze, the use of stamp seals, and the idea of writing and pictograms all suggest that the Indus civilization may have ultimately derived from Sumer; but links were neither continuous nor very strong. Indus metal tools developed independently, and whereas in Sumer there were early technical advances such as the use of a two-piece mould in casting, in the Indus cities use of the open mould continued much longer. The still undeciphered script, too, seems quite different from those of the Near East.

The Indus valley cities were not able to found a long tradition, like those of Egypt or Sumer, for about 1500 BC their ordered societies were brought to a violent end. Indo-European peoples, with chariots, stormed down from the north-west, burning and looting their way through the Indus valley and leaving a trail of devastation. The cities of the Indus disappeared; their mention in the *Rig Veda*, the oldest Indian literature, was thought to be legend. But their rediscovery and excavation has brought mythology into the realms of history.

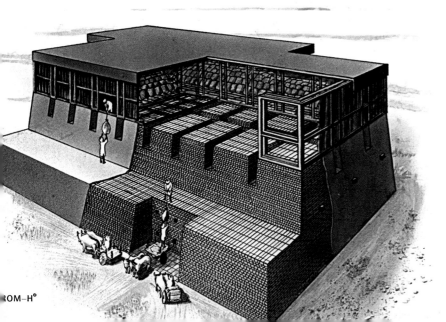

ROM-H°

Europe before the Romans

In the Near East the first cities had been established by 3000 BC; their complex societies using metals and the potter's wheel were based on a farming surplus, and had contacts with distant communities through trade. Conquest led to the forming of larger and larger units which formed the basis of future empires.

But in Europe (except on the fertile island of Crete) development took a different and poorer form. For here the environment did not often allow the concentration of resources necessary for the establishment of large centres of population. The prosperous Bronze Age cultures of south-east Europe, where cities might have developed, were blotted out by large-scale raids about 2000 BC and lapsed into simple village life.

The late third and the second millenium BC saw the regular use of gold, copper, and bronze in Europe and the rise of small, militaristic 'chiefdoms' either near the sources of metals or on important trade routes. Sea power brought prosperity; first the Minoans of Crete and then the Mycenaeans of mainland Greece dominated the Mediterranean trade before supremacy passed out of European hands to the Phoenicians. To the north and west, rich graves and buried treasures attest the existence of powerful rulers in Hungary, south Germany, Czechoslovakia, Iberia, Denmark, Brittany, and parts of Britain. But these were not city civilizations – rather they were 'heroic societies', led by kings or chiefs who probably claimed descent from gods as in later times. With them, in defended citadels or hill forts, lived their warriors who were rewarded for their loyalty by gifts of splendid weapons and ornaments of gold and silver,

Above: A fragment of a fresco found at Knossos, showing a lady of the court. Below: An aerial view of Knossos; Sir Arthur Evans, who was in charge of its excavation, restored several of the buildings with their characteristic deep red pillars. The whole hill must have been covered with such structures. Beneath the palace were drains, some large enough for a man to crawl along. The restorations indicate a palace of great beauty and luxury.

A gold ring and a bull's-head rhyton or ritual vessel, both from Minoan Crete. The rhyton is made of steatite with onyx eyes and horns of gilded wood.

made by craftsmen who possibly travelled from one centre to another, since the technology of this 'high Bronze Age' and later shows much interchange of techniques and art forms between the different parts of Europe.

It was not until late in the first millenium BC that true civilization spread with the rise of cities of the Etruscans in Italy, the classical Greeks, and later of the Romans, whose empire extended city life throughout Europe from the Danube and the Rhine to the north of England.

Left: A Minoan jar from a tomb just north of Knossos; 16 centimetres high, it dates from the 14th century BC. Minoan potters made vessels of many different forms, using the wheel; many were decorated with flowers or sea creatures, like this one.

Below: The throne room at Knossos; the frescoes are restorations. This room dates from the final period of the palace after the invasion from the Greek mainland in 1450 BC; the frescoes are of the Mycenaean type.

The Minoans of Crete

Crete lies in the southern Mediterranean between mainland Greece and Egypt. It is a fertile and well-watered island and a flourishing neolithic culture developed there. From this there emerged about 2000 BC a highly organized civilization based on large palace complexes, quite different from the civilizations of the Near East with their overpowering and conventionalized cities. This Cretan civilization was named *Minoan* by Sir Arthur Evans, who began excavations at Knossos in 1899 and identified it with the seat of King Minos of Greek legend.

On the fertile Cretan plains farmers were able to grow a huge surplus of corn and other crops that was used to support non-producers – kings, officials, priests, craftsmen, and traders. Wine, raisins, olives, olive oil, and timber were exported to Egypt and other east Mediterranean lands; Minoans with characteristic pots

are recognizable on Egyptian wall-paintings. Gold and stone for hollow vases returned to Crete from Egypt, whose art influenced that of the Minoans in several ways, including motifs of sphinxes and the convention of colouring men red and women white in wall-paintings.

The Minoans lived in isolated farms, villages, cities such as Knossos with its great palace, and small towns such as Gournia, a tightly packed conglomeration of five- or six-roomed houses clustered round a small palace. All the palaces were built to a uniform plan round a rectangular central court, which was unroofed and orientated north to south. From it, wide staircases led to upper floors.

Many rooms in the palaces were used for storing produce; it is estimated that over 1 million litres (about 220,000 gallons) of olive oil could have been stored at Knossos in row upon row of huge jars. But in spite of the concentration of supplies and riches in the palaces there were no fortifications. Possibly the Minoans trusted in their fleet to keep invaders at bay.

The many small shrines of Minoan religion contrast with the great temples of the contemporary Near East. Some were

within the palaces, but others were in caves where numerous votive offerings were left. Scenes on seals suggest that there were also small outdoor shrines, with a pillar or sacred tree. Sacred symbols like the double axe and 'horns of consecration' were always associated with the Cretan Great Goddess. A male deity rarely appears, though a young god, always smaller and subordinate to the goddess, may have been her consort. The Cretans buried their dead in caves or in collective tombs above or below ground level; there are no royal tombs with rich grave goods.

The greatest achievements of Minoan art and influence have been dated from 1700 to 1450 BC. Colonies were sent out to Aegean islands, including Thera (Santorin) and possibly to mainland Greece, as pottery and weapons found there, including those at Mycenae, were greatly influenced by Minoan forms if not of Minoan manufacture. But at the end of the 16th century BC the volcanic island of Thera blew up and much of it sank below the sea (possibly the origin of the Atlantis legend). In Crete the sun may have been blotted out for days by ash and tidal waves may have flooded towns and villages along the north coast. The palaces survived the chaos of the eruption, but about 1450 came disaster in the form of invasion from the Greek

Top and above: Clay figurines found in a temple at Mycenae during excavations of 1968–1969. Model snakes like this had not previously been found in the Aegean; some 17 were found in a store, some with separate tongues. The head of a female idol, 30 centimetres high, was one of a group of 19; they had bodies of coiled clay and the heads were separately shaped. This idol has spirals on the forehead, and three long tresses of twisted hair; the small holes at the base of the neck may have been for the attachment of ornaments.

mainland. Towns were destroyed by fire and palaces abandoned except for Knossos which suffered only superficial damage, as if the invaders preserved it to give prestige to their own rule. From this time, Crete was dominated by Mycenaean Greece.

The Mycenaeans

Archaeological evidence suggests that the Mycenaeans were an Indo-European people who came into Greece around 2000 to 1700 BC, probably from Iran via northern Anatolia. Their distinctive grey wheel-made pottery imitated silver vessels. The decipherment of their 'Linear B' clay tablets has demonstrated that they spoke a form of Greek. Some were warriors with horse-drawn chariots; others were farmers who grew corn and kept sheep, goats, pigs, and some cattle.

The Mycenaeans did not achieve a truly civilized society as their environment did not allow the rise of large concentrations of population in cities. Greece lacked the rich natural resources of Crete, and was divided by barren mountain ranges that made communication easier by sea than by land. The seafaring skill of the Mycenaeans is demonstrated by their many colonies, and by the incredible uniformity of culture – embracing architecture, religion, burial, dress, and writing – that they established over the Aegean area and Cyprus.

Small Mycenaean settlements grew up at strategic points, such as Mycenae itself, each dominated by a heavily fortified citadel. Local chiefs owed allegiance (according to later Greek tradition) to the high-king of Mycenae. On smaller hills around the citadels were tribal villages.

Within the citadel, the chief lived in a rectangular *megaron*, a distinctive plan also found in western Anatolia. It consisted of a pillared portico, an antechamber, and a hall with a great central hearth. Around the megaron were the domestic quarters – kitchens, storerooms, and the queen's apartments.

Unlike Minoan palaces, Mycenaean citadels were strongly defended, with walls of vast 'Cyclopean' blocks of stone. At Mycenae itself water could be reached in the event of a siege by long secret staircases in the thick of the walls.

Evidence for Mycenaean religion comes from finds of figurines, scenes on seal-

THE DECIPHERING OF LINEAR B

During his excavations at Knossos, Sir Arthur Evans found some 4000 tablets made of sun-dried clay. He studied them and recognized certain signs as portraying horses, chariots, and weapons. He established that the first Cretan writing consisted of hieroglyphs on gems and seals, and on a few tablets, dated to just after 2000 BC. From these a cursive form of writing developed that Evans called Linear A, which was used between the 18th and 15th centuries. (This is still undeciphered.) A more advanced script, Linear B, was evidently related to Linear A, and this form of writing was found on most of the Knossos tablets and on tablets found at Mycenean sites on mainland Greece as well.

The Linear B tablets were finally deciphered in 1952 by a British architect called Michael Ventris, assisted by John Chadwick. Ventris had been enthralled when, as a schoolboy, he heard Sir Arthur Evans lecture about the mysterious script; during the Second World War he worked on code-breaking, which gave him invaluable experience. Earlier scholars had classified all known signs, and Evans himself had shown that many of the signs were probably syllabic. Ventris managed to identify the place names 'Ko-no-so' (Knossos) and 'A-mi-ni-so' (Amnisos), with the help of the 'Classical Cypriot' script which, although much later, was clearly related and had seven similar signs. But Ventris was obsessed with the idea that the Linear B tablets were written in the lost Etruscan language and could make no further progress. At last, simply as what he called a 'frivolous digression', he tried giving Greek values to the signs. To his amazement he found that he could make sense of the tablets, allowing for the Greek being a very archaic form compared with Classical Greek.

One of the tablets with an inscription in Linear B found at Pylos.

The shaft graves at Mycenae. A circular double wall of limestone slabs, originally linked with horizontal slabs, surrounded Grave Circle A, which was excavated by Heinrich Schliemann in 1876. Within this circle were six shaft graves some 4 metres deep, with two to five skeletons in each accompanied by rich grave goods; these dated the burials to the later 16th century BC. Slightly older burials, from the beginning of the 16th century, were found in Grave Circle B, which contained 24 tombs, six of which were shaft graves. These were excavated from 1951 to 1954. The shaft graves seem to be copied from the tradition of royal burials of the Near East, such as Alaça Hüyük and the death pit of Ur. Over the faces of some of the dead were death masks of gold like the one below. This, with traces of attempts at mummification, suggests links with Egypt which may have been the source of the gold. In the graves were gold and silver vessels, crowns, and other ornaments, including splendid weapons much influenced by Crete, and possibly made by Cretan craftsmen.

Below: A gold rhyton in the form of a lion's head, from Shaft Grave IV at Mycenae – the richest of all the graves. It was hammered into shape from a single sheet of gold.

stones, and from the Linear B tablets at Pylos, which list offerings to different deities. The names of the 12 Olympian gods and goddesses of later Greece appear in the tablets, possibly arising as different aspects of the Cretan Great Goddess and her male consort. But in contrast to Minoan religion, the male element dominates the female, the god Poseidon, whose name means 'earth-husband', being the most revered of all deities. Unlike the Minoans, the Mycenaeans made life-size statues of their gods. But despite such divergencies the Mycenaean religion was derived from the Minoan, sharing its preference for small shrines, its use of identical ritual vessels and symbols such as the double axe, and reverence for the snake. Mycenaean deities and cult centres survived into Classical Greek times and the four great Greek sanctuaries of Delphi, Olympia, Eleusis, and Delos all had Mycenaean origins.

Mycenaean Greece reached the height of its power between 1500 and 1300 BC, but its decline after this was accelerated by the Trojan War which possibly took place about 1250 BC. About 1200 BC Mycenaean Greece suffered a crushing blow when a great raid from Central Europe devas-tated the countryside and destroyed many of the palaces, causing the virtual depopulation of much of central Greece. Refugees carried on a late flowering of Mycenaean culture in Cyprus and some centres, such as Athens, survived in spite of a total breakdown of communications and the collapse of Mycenaean unity.

During the 11th century BC Greece was peacefully reoccupied by Dorians from the north, who introduced the use of iron. After a 'Dark Age' that followed the collapse of the Mycenaean empire, Athens, about 900 BC, led a resurgence of city-states that culminated in the achievements of Classical Greece in the 5th century BC. The great days of the Mycenaean past survived only in tales gathered together as the basis of the Iliad by Homer in the 8th century BC, until in 1876 the excavations of Schliemann turned legend into history.

Indo-Europeans

Languages spoken in Europe today belong, with some exceptions, to the Indo-European group that includes Latin, Greek, Sanskrit, and Celtic. The Indo-European peoples have been traced to a homeland probably in South Russia, from

STONEHENGE

In Britain, the beginning of the second millenium was characterized by the construction of vast 'henge-monuments' – circles of huge tree-trunks or of stones, surrounded by a bank and ditch. The great ditch at Avebury in Wiltshire was 6 metres (20 feet) deep, hacked out of the chalk with deer-antler picks, as metal was not used to make digging tools. Avebury, which had two small stone settings inside the great outer circle and its avenue, was succeeded by Stonehenge, which is unique in the careful dressing of the surfaces of its stones. It has been claimed that the henges, the later stone circles, and stone alignments were erected with great precision for astronomical sightings. Such mathematical precision seems, however, too sophisticated for the simple material culture revealed through finds from burials. The henge tradition ended about 1400 BC, possibly through climatic deterioration.

Stonehenge in south-west England, from the air.

which, about 2000 BC, they began to migrate to east and west. In the east, they spread devastation in the Indus civilization about 1500 BC. In the west, they moved through Bulgaria into Anatolia about 2000 BC, leaving a trail of destruction – 300 burned sites dating from this time have been found in Anatolia. One group occupied Greece, and gave rise to the Mycenaean culture. Others can be identified in many areas of Europe as the Battle Axe, Globular Amphorae, Single Grave, and Corded Ware cultures, distinguished by beautifully shaped stone shaft-hole axes, and by pottery decorated by impressions left by lengths of cord pressed into the clay before firing. Their use of metal was minimal. The dead were buried in individual graves beneath a round burial mound or barrow. In central Europe they mingled with the Beaker Folk, a

Right: Three chalk cylinders, carved with a flint tool into designs that include the 'owl' or 'eye' motif, 90 centimetres high. They were found in the grave of a child of five years old in a round barrow on Folkton Wold in Yorkshire. The smallest was at the child's head, the other two at its hips. The designs are similar to those found in Ireland and Iberia.

This bronze sun-car from Trundholm in Denmark was probably made in central Europe in the 13th century BC. It is 60 centimetres long. The horse was cast by the lost-wax process; the disc, one side of which is covered with gold leaf, consists of two separate pieces joined at the edges by a bronze ring. The sun-car may represent the sun being drawn through the sky; it was found in a peat bog in 1902 and was probably thrown into a lake as an offering.

people named after their distinctive drinking cups.

In the Iberian peninsula (Spain and Portugal) and in the south of France a number of fortified settlements have been dated to about 2500 BC. Los Millares, in southern Spain, was defended by a bastioned wall. Its cemetery of great passage graves represented enormous communal effort. These settlements have long been regarded as 'colonies' from the Aegean, exploiting the rich metal resources of the west, but it is possible that they were due to indigenous development.

Central Europe

In Central Europe, deposits of copper and tin led to the rise of prosperous bronze-using communities during the second millenium BC. This was later the homeland of the Celts, and these earlier cultures, together with the older Beaker Folk from whom the population must have derived, may have been ancestral to the Celts.

In South Germany and Czechoslovakia a great tradition of metalworking of

increasing ability can be traced in the successive Únétice-Tumulus-Urnfield cultures beginning around 1800 BC. Through this area passed the great Baltic-Mediterranean trade routes, and different forms of axes and daggers were widely traded between centres in central and northern Europe, Italy, and the British Isles. Farming must have been very efficient, as vast numbers of burials in cemeteries suggest a large population which, from the 13th century on, began a series of migrations to east, west, and south.

Urnfield people
From the 13th century on the rite of cremation became fashionable in central Europe, the ashes of the dead being placed inside distinctive burial urns and interred in cemeteries. Finds of these urnfields from the end of the second millenium and the beginning of the first show migrations of these central Europeans into France, Iberia, and into Italy where the Urnfield population gave rise to the Villanovan culture, which developed about 900 BC into the Etruscan. The Urnfield people invented the brooch (anticipating the modern safety-pin) and a heavy slashing sword. They may have been responsible for the great raid that devastated the centres of Mycenaean Greece about 1200 BC. It is possible that their advanced technology was due at least in part to captured Mycenaean smiths who taught native smiths the art of making beaten bronze helmets, armour, shields, and large vessels for serving wine. It may have been at this time that two-wheeled war chariots were introduced to central Europe.

Meanwhile in Hungary and Romania settlements were established and occupied for hundreds of years forming great mounds or tells, such as Toszeg in Hungary. Smiths made distinctive axes and swords with blades decorated with spirals, and golden ornaments. Ceremonial objects include silver battle axes, and a great dagger of 21-carat gold found at Perșinari in Romania which even incomplete weighs 1·4 kilogrammes (3 pounds). Pottery was sophisticated in form; female figurines suggest that women wore dresses embroidered with patterns similar to those still found in the elaborate embroideries of the region. In the 13th century BC these tells were deserted, perhaps due as much to

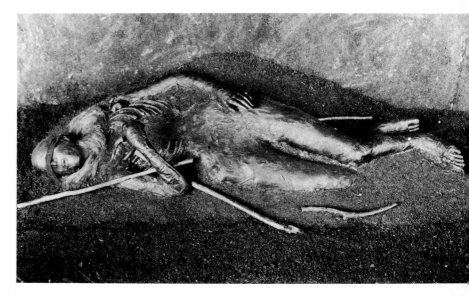

THE THRACIANS
The Thracians, a people whose origin is so far unknown, occupied much of south-eastern Europe from Macedonia to the Danube. They have been identified archaeologically as early as 1600 BC. This area was open to influences from four great cultures during the first millenium BC: the Greeks, whose colonies on the Black Sea coasts were founded in the later 7th century BC; the Celts to the west; the Scythians of the steppes to the north-east; and the Persians who were established in Asia Minor in the 6th century BC, and who occupied parts of Thrace for 30 years after the campaign of Darius in 512 BC.

Thracian craftsmen absorbed elements from the art of all these cultures, especially oriental motifs from Persian metal vessels and embroidered robes that fell into Thracian hands as gifts and loot. There were rich mineral deposits in Thrace, and craftsmen produced wonderful works of art for their royal masters. Gold and silver vessels, armour, and jewellery have been found in hoards and burials of the first millenium BC.

The Greeks regarded the Thracians as a mysterious people, and the priest-king and musician Orpheus gained a place in Greek legends. The 5th-century traveller Herodotus described Thracian life. Men preferred fighting to farming, which was despised as a way of life. Warriors were tattooed and fought in chariots, in which some were buried. Men had a number of wives, who worked hard both indoors and on the land. Death was welcomed as the next world was regarded as a paradise, while a new-born baby was greeted with lamentations, because of the long hard life ahead.

Above: The body of a girl of about 14 years old preserved for 2000 years in peat at Domlandsmoor in north Germany. She had been blindfolded, the left side of her head had been shaved, and she had been thrown into a lake to drown. This was the fate of unfaithful wives as described by the Roman writer Tacitus. Over a hundred similar 'bog-bodies' have been found in Denmark and north Germany.

Above: A Thracian silver-gilt plaque, of the 4th century BC; it is 5 centimetres high. It shows a horseman attacked by a bear. He is holding the reins in his left hand and a spear in his right, and wears a cuirass and greaves. Under the horse's hooves is a wolf.

A Thracian bronze matrix used to impress a design on metal beakers; it is 29 centimetres long. The design shows the influence of Scythian and Persian art.

119

climatic conditions as to the incursions of Urnfield warriors from the west.

Northern Europe lagged far behind the southern and central areas in its use of metal, and only round this time – 1200 BC – did the Northern Bronze Age begin in Denmark and north Germany. This region had to import all metals, exporting in return amber, a fossil resin washed up on the beaches from ancient forests submerged beneath the North Sea. Through the custom of throwing offerings into lakes, a number of ritual objects have been found preserved in peat.

The coming of iron

The Urnfield peoples lived in defended settlements, based on animal or cereal farming. The evidence of place names strongly suggests that they can be regarded as Celts. Miners and smiths were important specialists, and increasing use of bronze and skill in its working spread through Europe. But about 700 BC a culture arose in Austria that used iron. It has been named after Hallstatt where a great cemetery of 2500 graves has been found. Although the Hallstatt smiths used beaten bronze to make ornaments and elaborate vessels, some of the swords buried with the warriors were made of iron. The wealth of the Hallstatt people came from the copper and salt which were extensively mined in the region. The Hallstatt culture developed from that of the older Urnfield peoples but new factors, as

Above: This bronze bowl with a cow and calf was found in the cemetery at Hallstatt in Austria; its use is unknown. The cow's front legs rest on a support rising from the base of the bowl. The geometric patterns around the top of the bowl are typical of Hallstatt art, and contrast with the flowing tendrils of the later La Tène art.

	SCANDINAVIA	BRITISH ISLES	CENTRAL & SE EUROPE	WESTERN EUROPE	CRETE, ITALY, GREECE
	Domlandsmoor Girl	Caesar's raids Belgae settle in SE England	Thracian treasures La Tène culture	Caesar's conquest of Gaul Great hill fort towns La Tène culture	Ascendancy of Rome: 200s BC
	First ironworking	Hill forts in Scotland	Great hill fort towns		
500		First ironworking		Massilia: 600 BC Urnfield and Hallstatt cultures	Athens; the Parthenon
			Hallstatt culture—first ironworking, copper and salt mining		Etruscan civilization
		Bronze in common use			
1000					Dorians in Greece
	Trundholm sun-car Tree coffin burials Bronze in common use		Urnfield culture in C Europe Tumulus culture in C Europe		End of Mycenaean civilization Linear B writing Eruption of Thera
		Stonehenge in present form	Thracians in SE Europe	Argar bronze-using culture in Spain	
1500		Standing stones			Late Minoan Crete
		Stone circles Round barrows First metallurgy	Bronze-using Unetice culture in C Europe		Mycenaean shaft graves
2000	Corded Ware pastoralists Flint mines	Beaker Folk Flint mines	Corded Ware pastoralists		
2500	Passage graves	Passage graves	Beaker Folk	Los Millares citadels	Palaces in Crete Middle Minoan Crete
					Early Minoan Crete
3000	First farmers Funnel beakers	First farmers—long barrows, pottery	Copper shaft-hole axes in Hungary	Mixed farming	

A bronze wine flagon from a Celtic burial at Basse Yutz in northern France, 39 centimetres high. It dates possibly from the early 4th century BC. The ornament includes the three separate elements that were skilfully interwoven in La Tène art: the oriental beast (handle), and human mask (base of handle); the Greek palmette design (below the spout), and the Hallstatt water-bird – the little duck which is sitting on the spout.

well as the use of iron, included inhumation – burial of bodies in the ground in timber-lined shafts beneath large barrows. Tribal centres were established in defended hill forts with timber-framed ramparts, as in Urnfield times. Richer and poorer graves show that society was divided into classes. Royal burials contain a four-wheeled waggon used as a hearse. The status of women was evidently high, since a woman buried at Vix in France in a timber-lined shaft beneath a large stone cairn was provided with Greek and Etruscan wine vessels and cups. On the woman's brow was a golden diadem ornamented with two little winged horses, possibly of Iberian manufacture. Greek influence and artefacts spread from the Greek colony at Massilia (Marseilles), founded about 600 BC, up the rivers Rhône and Saône to the Upper Danube region.

The La Tène culture

About 450 BC in the Rhineland the La Tène culture developed from the Hallstatt. It is distinguished by its sophisticated and original art on metalwork and pottery. This must have been stimulated by Greek objects from Massilia, and by Etruscan and Greek vessels brought across the Alps. The art arose suddenly, craftsmen brilliantly combining the Hallstatt sacred waterbird with Greek motifs such as plant tendrils and the palmette, and also with grotesque animal and human heads derived from the art of the Scythians (nomads of the Eurasiatic steppes who produced complicated and beautiful metalwork) and Persians through eastern Europe and Italy.

The La Tène culture was warlike, the art being found on helmets, shields, and scabbards, in addition to splendid golden torcs and armlets. In the last centuries BC the La Tène Celts migrated southward to attack the Greeks and Romans in their homelands. During the 1st century BC the Romans brought much of western Europe into their empire. They imposed their highly structured civilization on the peoples who lived there, and their mass-produced artefacts brought individual Celtic craftsmanship to an end. Celtic art continued in Britain until the Claudian conquest of the 1st century AD, and in Ireland which was never conquered by the Romans the La Tène culture persisted into Christian times, when its ancient art motifs were used to illuminate the manuscripts of the monks.

Above: A Celtic bronze shield found in the river Thames at Battersea, London, possibly dating from the 1st century AD. It is 77 centimetres long. It would have been fastened to a wooden backing. The designs include four owl faces surrounding the central roundel, with a stag's head linking the central circle to the circles at the top and bottom.
Below: This silver-gilt bowl from Gundestrup in Denmark is probably Thracian work but shows Celtic influence. It is 69 centimetres in diameter. Twelve separate plaques portraying gods and ritual scenes are fastened to its top.

Early Asia

By the end of the last Ice Age about 10,000 years ago, man's environment had altered greatly. Not only had plants and animals changed with the warmer climate, but many outlying parts of the Asian land mass had been turned by the rising sea level into islands.

In some ways the life of early men in the Far East seems to have continued unchanged after the Pleistocene, judging from the artefacts which have been unearthed. But after this period there appear the remains of meals with components unlike those seen previously, new types of artefacts, or the presence of other items which seem to indicate man's adaptation to a somewhat changed environment, or a progression from more basic techniques to those implying refinements of earlier skills.

Study of ancient pollen from lake sediments in Taiwan, and of animal bones from postglacial geological deposits and archaeological sites in China, point to a gradual rise in temperature, accompanied by an increase in vegetation, the extinction of such cold-climate mammals as the mammoth and woolly rhinoceros (in North China), and the appearance in various areas of warm-climate species such as rhinoceros and water buffalo.

At this time the inhabitants of the Mongolian steppes manufactured small stone tools known as Gobi microliths, which have been found in the sand dunes of ancient oases. These tiny tools are quite distinct from those typical of Western Asia, and may have been set in shafts of wood or bone for use as spears. In the advanced stages of this phase, pottery decorated with geometric or incised patterns and cordmarking was used. Ostrich eggshells were found fashioned into rings at Mongolian sites, and these eggs may have been a dietary staple.

In North China and Manchuria, the forest-dwellers expanded on many of the traditions of their predecessors. The Upper Cave of the Choukoutien site (see page 64) still yields the bones of hyaenas, bears, and elephants from this date, but for the first time the remains of strictly

Bronze-working appears suddenly in China during the Shang dynasty (1600–1027 BC). The Shang craftsmen possessed many advanced techniques; some of their finest work is seen in sacrificial vessels. These were of traditional shape according to function, and were buried with important people. The vessel above, called a ting, was a container for food offerings, possibly grain. It was made in a mould of several pieces; the flanges at the corners make a decorative feature of the projections caused when molten bronze flowed into the joints.

modern red deer also occur. These people continued the palaeolithic stone tool forms found at the site such as choppers, scrapers, and flakes, though bone and antler were now used as well, and were even employed in fashioning needles. Perforated teeth of tigers, foxes, deer, and badgers were strung with fish bones and mollusc shells into necklaces. Sea shells at this inland site are important indications of either long-distance travel, or trade with the coastal area hundreds of kilometres away. Human skeletal remains were found in this Upper Cave, with skull injuries which might have been the result of violence. The surrounding earth had been strewn with red ochre, apparently as part of the funeral ritual; this again had been a practice of their ancestors. At other sites in the region, axes suitable for tree-felling were used, while along the sea coast tools such as arrowheads and awls were sometimes made of bone, and coarse pottery was produced.

In South China during this early postglacial period typical finds are chipped stone tools similar to the old 'chopper' style, animal bones, seeds of wild plants, and mollusc shells. In the forests of the

Above: Three arrowheads used by the Jomon hunters of Japan. These were generally made of stone, although bone was sometimes used; the finest were made of the volcanic glass, obsidian.

south-west, canoes may have been made from tree-trunks with the aid of heavy stone axes.

These pre-agricultural groups in China then were hunter-fisher-gatherers, living in seemingly impermanent settlements in the open or in caves close by rivers, streams, or the sea, which were an important source of food in the warmer climate after the glacial period. They used stone – flaked and at times polished – bone, and antler as well as wood for implements; and some groups were beginning to manufacture pottery.

Island cultures

Once the glaciers had melted, the islands which make up Japan were fully separated from the Asian land mass since the regions which had previously been land-bridges were now submerged. The first post-Pleistocene cultures of Japan are generally considered to be *Jomon* ('cord-pattern'), called after their early cord-wrapped pottery dating from over 9000 years ago, as at Natsushima shell-mound near the mouth of Tokyo Bay. Still older ceramics (known as *Ryutaimon* or 'raised-band pattern') from Fukui Cave in Kyushu have been dated by radiocarbon as more than 12,000 years old, which makes Japanese pottery the oldest yet discovered anywhere in the world! As even this is not primitive in form, it is clear that the origins of this pottery-making are of even greater antiquity, either in Japan itself or an adjacent area. The fact that the appearance of this Ryutaimon pottery is different from that usually considered typical of Jomon has led to some controversy as to whether it is in fact part of the Jomon tradition, or of a different one.

The Jomon cultural tradition had several stages and lasted some 8000 years. Traits include shell middens (see page 82), thousands of which have been discovered; semi-subterranean dwellings (probably an accommodation to the cold climate in certain zones, and sometimes equipped with fireplaces and paving stones); stone implements (such as hunting points, chopping tools, and chipped or ground axes, and scrapers useful for cutting hides); and pottery with cord markings or various other forms of decoration such as finger-nail markings, shell impressions, or zigzag patterns.

Below: The Jomon people lived in rectangular or round pit dwellings like these, constructed of uprights and beams covered with bark or leaves. Villages of 15 or more are often found. The huts are sometimes floored with stone.

Evidence has been found in Japan for deep-sea fishing which implies seagoing vessels even 9000 years ago. Stone and clay sinkers and bone fish hooks were used in fishing. Deer and wild boar supplied meat, marrow, fat for oil, and skins. Adornments such as ear-rings, bracelets, and necklaces were made of shells, horn, stone, clay, and animal teeth. Clay figurines depicting humans and animals were kept in many households, and perhaps were related to fertility, or had other spiritual or magical significance. There was a variety of burial practices, from bodies curled up or stretched out, to burial in jars, the use of red ochre on the skeleton, or covering the body with stones.

Isolated groups

South-east Asia extends from Burma, Thailand, Cambodia, Laos and Vietnam through Malaya, the Philippines, and the Indonesian islands, and eastward to include Australia. The postglacial rise in

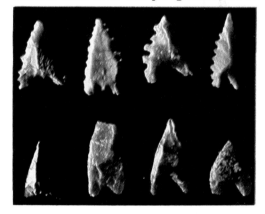

Above: These small, hollow-based stone points, known as Maros points, are common in cave deposits from Sulawesi, Indonesia. The saw-toothed margins make a deep wound and possibly helped to spread poison through the victim's body. They are about 4000 years old. Below: A Jomon pottery figure.

sea level cut off many previously connected land masses from one another, and as a result some plants and animals never reached certain of these areas. This 'trapping' of some groups of plants and animals led to different patterns of feeding and of cultural development, at least until man developed sea transport, either of simple rafts or canoes of bark or tree-trunks, for travelling between the islands.

These lands, like most other regions at this period, were peopled by gatherers, fishermen, and hunters, living near water and probably moving frequently according to the availability of food. Where the food supply was relatively constant, as near coastal shellbanks, settlements may have been somewhat more stable. Besides the remains of meals of shellfish left behind in shell middens, diet here included various game animals, especially pig and deer, edible roots and tubers (such as yams), and berries and nuts. Shelter was provided by caves, or possibly simple structures of wood, bamboo, and leaves. Bamboo also could have been used with moss to make fire by friction, while other plant material so plentiful in the humid tropical climate would have been useful for basketry and fish nets. Hunting tools may have been mainly of wood or bamboo, and have included blow-pipes, spears, and poisoned arrows. Stone tools of the *Hoabinhian* type (named after an archaeological site in Vietnam) were made from river pebbles and consisted of choppers and other artefacts worked on one side, flaked tools, axes, and grinding stones, and were part of a long-lasting and widespread tradition throughout much of South-east Asia; although most of the islands of the archipelago had diverse flake and blade stone industries instead. Cord-marked pottery

In the coastal regions of South-east Asia, prehistoric peoples who lived mainly on shellfish have left vast rubbish-heaps, or shell middens. These have since been covered by layers of soil. They are a rich source of information for the archaeologist – but they are also rich in calcium carbonate, used in the manufacture of lime. Many, like this one at Hinai, North Sumatra, are excavated for commercial, not scientific, purposes.

Below: A selection of neolithic stone adzes, found in Java. They are made of local stone, highly polished for efficiency and beauty.

was commonly used. Though many areas of South-east Asia shared these common traits, local industries sometimes varied, as in some of the islands which remained more isolated.

In some areas cave or rock paintings of animals and symbols are found. Stencils of human hands, even indicating missing fingers, are portrayed too, and resemble those found in European caves. It is difficult to determine whether this was a result of mutilation, or merely of bending backwards of the fingers. While some of the paintings show items familiar from man's environment at the time, others perhaps served some unknown ceremonial function, or may have simply been drawn much as 'doodles' of today are.

Australia

Australia, at the south-east limit of the area, had relatively scarce wood resources, and was lacking in bamboo. This is important in understanding the continuity and specialization in stone tools found here. Edge-ground axes are found in the Australian tool-kit from well back into the Pleistocene; much later, leaf-shaped points and blades made their appearance along with other types. Some features, such as hafting of tools, were also employed in New Guinea; these may well have been the result of parallel developments, and not necessarily the result of migrations, though in some aspects of their culture the islands still farther east in the South Pacific are indeed thought to have been influenced by their neighbours.

South-east Asia! If this is so, and therefore many major aspects of South-east Asian culture are indeed native developments, instead of merely ideas or practices 'borrowed' from other regions, it will be necessary to alter some of our previous thoughts on the rise, spread, and relations of these different early civilizations.

The site which is shaking up old theories is Spirit Cave, high on the face of a limestone cliff in north-west Thailand. The most startling finds here were fragmentary remains of numerous plant groups including pepper, candlenut, betel, cucumber, bottle gourd, water chestnut, peas, and beans, dating from 12,000 years ago – thus by far pre-dating such agricultural evidence from even the Near East. These species could have been utilized as spices (pepper), stimulants (betel), and for

Excavations in Australia and Tasmania indicate that the inhabitants caught fish, birds, and marsupials such as kangaroo, and that the dingo or wild dog played the role of companion or scavenger. The hunting, fishing, and gathering economy persisted in Australia right up until the arrival of Europeans and indeed is still found in modern aboriginal populations. Evidently this form of subsistence was indeed workable and Australians adapted successfully to their environment without ever incorporating agriculture or the use of pottery or metal.

The beginnings of agriculture

The introduction of agriculture generally marks an important milestone in the course of the development of societies, with the tending of crops in a specific area and the resulting adoption of a more sedentary way of life. Surplus crops stored for further use release the people from the hand-to-mouth existence which is often a feature of hunter-gatherer groups, and generally lead to increased population growth, and greater specialization of activities.

It has long been believed that agriculture (and indeed many other significant techniques of advanced societies) began and spread from the great civilizations of the Near East, India, and China, but exciting archaeological discoveries in Thailand have shown that the earliest known farmers anywhere may have lived there, in

Above: Some Aborigine groups in Australia still continue the hunting and gathering way of life. This man, pictured in New South Wales, is fishing with a spear, tipped with a double stone point. Right: Message sticks, used by Aborigines, serve to identify a messenger to other tribes and show that his errand is peaceful; sometimes they remind the messenger of details – so the number of notches on a stick might correspond to the number of days before a feast. Below: The skeleton of a child, buried in a pit over 20,000 years ago. The grave was one of many found on the 'shore' of an extinct lake in New South Wales.

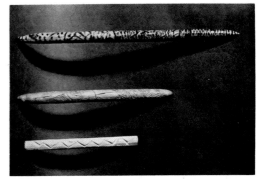

lighting (candlenut); indeed some are still used as such today. Though it is possible these plant remains might have been collected rather than grown, some botanists think they suggest actual horticultural

Right: The early island peoples of South-east Asia did not weave, but made cloth (above) by beating fibrous bark with ridged beaters, until it was thin and flexible.

Kaolin clay was first used by the potters of the Lung-shan culture in China, for pots like this k'uei tripod of about 2000 BC. Their experiments with this hard white paste led eventually to the development of porcelain.

Right: Ban Chieng, a burial mound in north-east Thailand, has yielded remains up to 5000 years old. Pottery from this site is usually painted or, like this example, decorated with incised patterns.

Below: A Lung-shan bowl of polished black ware. Such pottery, often wheel-thrown, was not glazed but fired by a special technique which resulted in a layer of carbon being formed on its surface.

activity. Later on in the Spirit Cave sequence are found new tool types such as rectangular, partially polished stone adzes, and slate knives resembling present-day Indonesian rice reapers, presumably evidence of a shift to agricultural activity. At Non Nok Tha ('Partridge Mound'), a site in north-east Thailand, the imprint of rice in pottery suggests the cultivation of rice dating back to 3500 BC or before, while bone remains at this site may be those of domestic cattle.

We do not know precisely when these traits moved farther south and then east into the islands of South-east Asia. The populations of these islands, while they continued to fish and hunt, learned to make pottery and to grow first root crops and tubers such as yam and taro, and then rice. They domesticated animals – predominantly pigs and chickens, and perhaps dogs. Fruits including the banana and sugar cane were later to prove important crops in the island regions. Stone bark-beaters from this period show that these people manufactured bark-cloth from specially softened fibres.

Chinese village farming

Agriculture in China evolved into an economy based on plant cultivation supplemented by animal husbandry. The earliest phases probably began with horticulture based on fruits, nuts, and roots, later expanding to the cultivation of the staples millet and rice, as well as vegetables such as cabbage.

The origins of village farming communities in China appear to have begun in the Hwang-ho (Yellow River) Valley, and this region is sometimes referred to as the North China Nuclear Area. This was a favourable area due to its position between wooded highlands and swampy lowlands, and its rich yellow loess soil.

The earliest well-established stage of these northern Chinese farmers has been named the Yang-shao culture (after a site in Honan) and first appeared more than 7000 years ago. Over a thousand sites have been attributed to this phase. The principal crop was the cereal millet; implements included hoes, spades, polished axes, and sickles of stone. Some of their distinctive painted pottery was used for grain storage, and grinding stones were used in the preparation of flour or processing of plants. The Yang-shao farmers domesticated dogs and pigs, and in some areas cattle, sheep, and goats. Silkworms may have been reared – a half-cut cocoon was unearthed at one site. Silk, hemp, and other fabrics were worked with pottery spindle whorls and bone needles. Numerous stone and bone points and arrowheads show that hunting continued, and a variety of wild animal bones (among them deer, antelope, and hare) have been recovered from refuse middens. The importance of fishing is shown by the many harpoons, fish-hooks and stone net-sinkers, and by fish motifs on pottery.

Pan-p'o, an advanced Yang-shao site in Shensi province, is the site of the most extensive remains of an early agricultural village in China. It dates from about 4000 BC. This spectacular settlement covers 70,000 square metres (17.3 acres) and included 46 houses, an area for burials and pottery kilns, and buildings believed to have been used to house domesticated animals. The houses had walls of earth and wood pillars bore the weight of roofs covered by clay and straw. Deep storage pits for grain surrounded the houses. The presence of many stone scrapers seems to indicate leather-working for garments, which were also made of woven fabrics. The dead were usually buried singly, some of them – probably important people – in wooden coffins. Three pots generally accompanied the corpse, and sometimes stone or shell ornaments were included. Children were buried in urns near the dwellings. Both at Pan-p'o and at other Yang-shao villages there is a large communal house, which may indicate co-operative farming efforts.

Yang-shao pottery was made by hand, and some finds indicate the coiling method. It ranged from thick-walled red cooking pots, coarse or fine red- and grey-ware for storage, cups for drinking, and fine polished and painted bowls. Basket-weaving must have been highly developed, judging by the mat-impressed decorations seen on the pottery at Pan-p'o.

In North China, the Yang-shao was replaced by the Lung-shan rice culture. This began about 3000 BC and is well known from sites extending eastward into Shantung province and along the eastern coast. These farmers also produced fine wheel-thrown pottery, characteristically black in colour. A typical form had a leg-ged tripod base, and made an efficient cauldron and steamer, enabling food to be cooked quickly. This cooking technique is still used in China today. The Lung-shan potters experimented with kaolin clay, paving the way for the later manufacture of white porcelain eventually known simply as 'china'.

Bone was still used for implements, and during the Lung-shan period a new use for bone emerged: the shoulder-blades (scapulae) of animals such as oxen and deer were heated, and prophesies made on the basis of the cracks formed. In the succeeding Shang dynasty this practice was developed with questions and answers being inscribed on the bones.

The advent of metallurgy

The beginnings of metal-working were dependent not simply upon locating suitable raw materials, but on the understanding of the processes by which they could either be beaten, or smelted and moulded or cast into shape as well. Prehistorians long assumed that the knowledge of these processes spread out from the highly developed and well known civilizations of the Near East to other, supposedly more backward areas. But it

Above: Fragments of oracle bones from the Shang dynasty. The pictographic script used in inscriptions on the bones survives, with many changes, to the present day. More than 10,000 oracle bones have been found at the ancient Shang capital, Anyang.

A red pottery bowl, found at Pan-P'o. The stylized face is unusual in Yang-shao decoration.

The Yang-shao farming village site of Pan-P'o housed two to three hundred people in dwelling huts like those reconstructed below, made with earthen walls, straw and clay covered roofs, and supported inside by wooden pillars. The site was divided into three distinct sections, a living area, an 'industrial' area devoted to the manufacture of pottery, and a burial ground. In the reconstructed interior of one of the houses, some items (like the red bottle hanging from the roof) were actually found at Pan-P'o; others, like the deer-skin and rush-matting are supposed to have been used on the evidence of other finds. Leather-working is indicated by small stone scrapers, and basketry in herringbone weaves by impressions on the bases of pots. Pan-P'o was excavated during the 1950s.

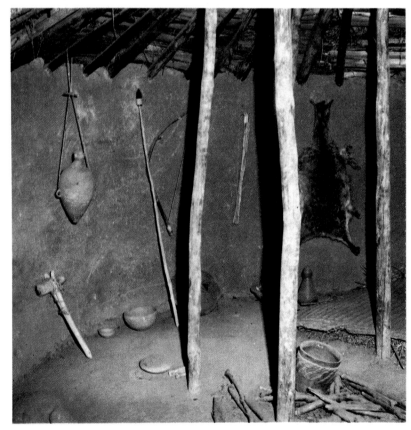

127

The oldest socketed metal tool in the world seems to be this one, found on the chest of a skeleton in a burial mound at Non Nok Tha in Thailand (below). Like the cord-marked pottery in the same grave, it is a funerary offering. The tool is thought to be 5500 years old.

now surprisingly seems that metalworking was probably *independently* invented in mainland South-east Asia. The evidence for this comes from Non Nok Tha in Thailand. Among the finds were sandstone moulds for casting, crucibles, and the bronze tools themselves, dated to about 2500 BC.

Even more astonishing was a copper socketed tool thought to be 5500 years old, found with one of the mound's 200 burials, which is not only the oldest known metal tool from eastern Asia, but the oldest socketed metal tool yet found anywhere. Neither China nor the Indus valley has yet yielded evidence of comparable metalworking at such an early date, while if the dating of the mound is accurate, it would not seem likely that techniques spread so

rapidly from the Near East, some 6000 kilometres (4000 miles) away (page 106).

Once men learned how to produce and work bronze alloys, a variety of local styles and decorations for metal implements developed throughout South-east Asia. Stone was often used side by side with the metal objects, which in the islands of South-east Asia are mainly cast, socketed bronze and copper axes and agricultural tools. Bronze was also used in this area for bells, bracelets, and occasionally for halberds, as well as for large, decorated kettledrums.

A puzzling aspect of these South-east Asian metal cultures is that they show little indication of urbanization or warfare. It seems that although plant and animal domestication was spreading and people engaged in metallurgy and trade, centralized political power was still absent. It was only around the time of Christ, notably under Indian influence, that any kind of centralized state was established; and after this the local cultures were dramatically transformed.

The Dong S'on necropolis

Excavations in North Vietnam in the 1920s revealed the necropolis of Dong S'on, in which the dead had been buried with numerous funeral deposits. Among

Above: Two decorated bronze ceremonial axes from Indonesia show the skill of the metalworkers of prehistoric South-east Asia. The socketed axe on the left is giant sized – about 70 centimetres long; the other, about half the size, is decorated with a bird carrying a similar axe.

THE KHMER CIVILIZATION

The civilizations of South-east Asia grew up under the influence of India or of China. But for some 400 years a powerful empire existed in what is now the Khmer Republic (Cambodia). And though it was clearly much influenced by India, it is also highly individualistic.

This civilization was founded by the Khmers in the 6th century AD; by the 9th century it was immensely powerful, and it lasted until the 15th century when it was overthrown by the neighbouring Thais. Its capital of Angkor in the northern jungle was destroyed and only rediscovered in the 19th century. Nearby was the largest religious building that has ever been known – the temple of Angkor Wat (left).

The Khmers were extraordinarily skilled and imaginative architects. Though they were Buddhists and their art was strongly influenced by India, Angkor nevertheless shows highly individual characteristics, among them the huge carved smiling faces, representing rulers raised to the rank of gods. Other typical features of Khmer architecture, which are not found in that of India, are half-vaulted arches and sloping roofs of curved stone.

the finds were bronze double-edged swords, socketed axes, spearheads and daggers, bells, buckets, and numerous kettledrums, some in miniature. The bronzes were decorated with intricate spirals and geometric designs. The dating of this site to a few centuries before the Christian era is based on Chinese objects in the burials, and indicates contact between the two areas.

In the Chinese province of Yunan, 700 kilometres (450 miles) away, the Late Bronze Age site of Shi-chai-shan on Lake Tien yielded the royal tombs of the king of Tien. The magnificent and ornately decorated bronzes found there are in some ways similar to the objects at Dong S'on, but the relationship between the two sites is not yet clear. Again there appear the large drums, although those from the Chinese site are more elaborate, with lively scenes and figurines on the drum tops. It has been suggested that these two sites were both influenced in their use of animal motifs by those of the Central Asian nomads; such a connection, however, remains controversial.

A late developer

Japan, cut off from the mainland, was slow to develop both agriculture and metallurgy. Grains such as rice and millet were

This large bronze kettledrum from Laos cannot be certainly dated, but its shape and method of casting were well known in prehistoric South-east Asia. Dozens have been found in the area, and some are still used on ceremonial occasions today. They were cast by an ingenious variation of the lost wax process (see page 106); rather than carve a mould with all the intricate bands of pattern on the side and top of the drum, the caster made a stamp of part of the pattern and pressed it repeatedly into a sheet of soft wax, which was bent to fit the core.

Right: A dōtaku, a bell-like object that is characteristic of the Bronze Age of Japan. Its use is unknown; it could have been a ritual musical instrument, its shape based on a Korean horsebell. Bronze reached Japan from Korea in the 1st century AD. Below: A cedarwood sword and bow of the Japanese Jomon culture. Both were preserved in peat bogs and retain traces of decorative red lacquer.

probably introduced there from the Chinese mainland; they do not seem to have grown in Japan until the latest Jomon period, about 1000 BC; indeed intensive agriculture, in the form of wet-rice paddy cultivation, did not begin until the succeeding Yayoi period. Hunting continued throughout the Jomon, as is witnessed by quantities of wild boar bones and the preservation in peat bogs of wooden, red-lacquered bows dating from the latest Jomon times. The bogs also yielded wicker objects, twisted nets, matting, wooden bowls and cups, wooden earrings, and a ceremonial wooden sword.

With the Yayoi culture came the first use of metal. The specialization involved in the production of the bronze- and ironware, wheel-made pottery, and rice farming indicates a stratified society. In the following centuries, contact with the mainland is shown by the graves of the wealthy, which contain jewellery, mirrors, and beads, many of which were imported from China. Halberd-blades, swords, and spearheads were typical weapons.

The shift to intensive rice agriculture at this time is indicated by the appearance in all major Yayoi sites of pottery with the imprints of rice, and stone rice reapers are also commonly found. The excavations of village sites in Japan indicated that dwellings were round or oblong, and that grain storehouses were constructed on supports several feet off the ground. At a waterlogged site, wooden objects included hoes and spades, household utensils, and wooden clogs essential for crossing the marshy fields. Oak and cedarwood were particularly popular. Looms were in use for weaving, and textiles, probably woven of hemp or ramie, have been recovered from burial jars. The dead were usually buried in pottery jars, in cist graves made of large stone slabs, or in jars under dolmens. The dolmens consisted of one large stone, supported by smaller ones in a circular formation. Some resemble sundials, and may be connected with worship of the sun goddess referred to in later writings.

The succeeding Tomb Period, which began in the early centuries AD, is one of imperial reign, with a complex society including aristocrats, petty nobles, and guilds of workers carrying hereditary positions: a strong clan system was a feature of the period. More than 10,000 tombs of the phase have been found, obviously requiring a large, mobilized work force acting for the benefit of the rulers. The large mounds often contain clay models (*haniwa*), frequently of houses, and provide details as to the structures of that time. The tombs of the 5th century are the largest, and a warrior class is evident from the armour, weapons, and horse gear. The period ended with the rise of Buddhism.

Bronze-Age China
The origins of bronze-working in China are still elusive. The earliest material known from the region is that of the Shang

The use of bronze ritual vessels continued in China for many centuries after the Shang. New shapes were introduced but older decorative motifs were often used. This tripod dates from the Spring and Autumn period (7th or 6th century BC; the deer shape is a departure from tradition but the coiled snake on its shoulder is found on many Shang bronzes, like the owl (top).

This Chinese bronze vessel in the shape of two owls back to back was used to hold sacrificial wine. It dates from the Shang dynasty, of the 14th to 11th centuries BC, and was cast in an elaborately carved multi-piece mould.

Below: A bronze ceremonial axe of the Shang dynasty, with openwork and relief decoration. The mask design is typical of the period.

dynasty, centred in the Anyang area, where sophisticated casting appears suddenly in about 1500 BC, with no more basic stage of metal-working yet discovered (see page 107). The skilled workmanship predates by centuries comparable material from Europe, and is equivalent to that of the Late Bronze Age in the Mediterranean. The craftsmen of the Shang period used motifs of monster masks, animal forms, and geometric patterns on lovely ritual vessels for food, wine, and water.

Under the Shang dynasty an organized and centralized empire grew up. Its immensely powerful ruler took all government decisions and acted as chief priest. The economy was still based on agriculture and the tools of the peasants hardly altered from those of the Stone Age, but the noble classes lived in comparative luxury. Excavations at Anyang, the chief Shang site, have revealed a well-planned city with a magnificent palace and royal tombs. Buried with the rulers were not only fine examples of bronze and pottery but animal and human sacrifices. Large numbers of chariots, and the skeletons of the horses that drew them, were also found. But our knowledge of the Shang dynasty comes not only from excavations; during this time the Chinese evolved a written pictographic script for inscriptions on bronzes and on oracle bones. This script formed the basis of that used in China and several other countries today.

Africa and its Past

Although the earlier African hominids had escaped the direct influence of the ice sheets which covered so much of the northern hemisphere during phases of the Upper Pleistocene period, they were nevertheless slow to seize on such advantages and push ahead with cultural advances. An important contributing factor may have been the fact that until about 35,000 years ago, much of Africa was peopled by the more primitive varieties of man (see page 69) – the Neanderthals in the north and Rhodesian people in more southern areas. Probably advanced varieties first became successful in more eastern regions; it is still a complete puzzle whether these hunters – the ancestors of many modern African tribes – evolved within Africa or came in from the Near East. Which ever it was, they eventually spread widely in Africa, replacing all the earlier men, and evolving into the distinctive African peoples known today.

As these advanced groups moved into new country, they brought with them well-tested knowledge and techniques. The challenges were greater for some groups than for others, and the environment was kinder to some than others. Highly complex societies evolved in the fertile regions of Egypt at a time when much of the rest of Africa was still poorly developed. It was much later that highly organized regional or tribal societies, the iron-working people of Nok in Nigeria and the builders of the awesome stonework and fortifications at Zimbabwe in Rhodesia, came into being and left archaeological evidence.

These late Pleistocene to more recent (Holocene) times saw great changes, especially the gradual reduction of hunting and gathering societies and the emergence of early farmers and metalworkers. Climate certainly influenced the well-being of these communities – it was

A prehistoric engraving of a cow, ostrich, and other animals from the Air Mountians of north central Niger, in West Africa.

cooler between about 15,000 and 10,000 years ago. In some areas south of the Sahara and in East Africa far more arid conditions then set in. This was followed by a change to a warmer and more humid weather which lasted until about 2500 BC. Further drying out took place later; the large East African lakes shrank in size and the area along the Nile grew more arid. Hand in hand with such changes went vegetational changes and a reduction in the numbers of game animals. The Sahara was probably at its best just about at the end of the Pleistocene, for both hunters and early pastoralists. The fact that Lake Chad was then about eight times its present size is some indication of the lush conditions prevailing.

Professor Desmond Clark, the prehistorian, believes that these environmental differences must be related generally to the cultural differences in these 'Later Stone Age' folk, but, of course, such earlier societies had their fads and fancies, as we do today. Minor differences in stone tools, or later in iron-working or farming methods, may therefore just indicate the inventiveness or preferences of regional groups. They can also indicate the continuation of different traditions in manufacturing, as we know from local evidence in the early Nile valley. Today we live in a consumer world flooded with new gadgets with a limited lifespan. Early communities at times had the wisdom to hold on to things which served them well enough. Occasionally, however, old-established ways seem to have been swept aside, as if by the excellence of a new idea. We see this in the spread throughout Africa of the microlithic tool-making methods during the close of the Pleistocene, in which small bladelets and flakes were mounted to produce a variety of new tool shapes, from multi-barbed spears and arrowheads to sickles and saw-blades. Evidence for their use appears from such widely separated areas as the Nile valley, the Congo, and

An example of early Bush-Hottentot cave art. The animals are sensitively painted and fairly naturalistic, while the humans are highly stylized.

the South African Highveld. Bone and ivory were also being used successfully, the sum total of all this being a noticeable overall advance in technology. In the case of one of these distinctive microlithic cultural groups, the Capsian people of Tunisia and Algeria in North Africa, there are vast amounts of associated food remains in the form of great shell mounds.

The Later Stone Age People

Evidence of the transitional period – between palaeolithic hunter and early farmer – is really very patchy for much of Africa. This period from about 6000 to 5000 BC was one where big game hunting was still important to some groups. In more northern parts, elephant, giant buffalo, rhino, large antelope, and hippopotamus were still important meat providers. The communities living in the more tropical forested areas were, of course, exploiting a very different environment, and like the Pygmies of today probably had to search the trees for smaller animals, although pigs could have been common in some areas. In the more southern areas of Africa, fairly open grasslands and thinly wooded savanna country provided both hunting and collecting opportunities. This, at least, seems to have been the case with the Nachikufan people of the Malawi region, who were trapping game in fairly open country. They also found it convenient to manufacture grinding stones to prepare some of the seeds or nuts they collected.

Some of the art of these Later Stone Age peoples has been preserved. This rock art takes many different forms. The oldest art work may be the engravings, markings hammered carefully into fairly smooth rock surfaces. Paintings occur both on exposed rocky areas and in rock shelters. Some examples reach a very high standard of artistry, with a careful use of different pigments and much effort and attention obviously being given to detail and proportions. As in the case of European cave art, a variety of animals is represented – rhino, warthog, eland, cattle, and even dog.

Like other prehistoric art, the African paintings and engravings were probably undertaken for a variety of reasons – to record events of importance to the hunters, in initiation rites, and as part of magic ceremonies. A rather different type of 'folk art' has been found in the form of bone and shell pendants and beads for adornment. Perhaps these people used bark cloth or skins and coloured them with pigments mixed in fat or resin.

These were clearly times of change, growing artistic feeling, and perhaps of increasing regional differences. Cultural and economic differences were more pronounced than ever before. From 5000 BC agriculture spread from the north, and similarly the use and value of metal became known south of the Sahara.

The emergence of farming
Exactly why certain plants started to be cultivated in certain parts of the world, and certain mammals domesticated, is still debatable. The factors probably included the problem of feeding ever-growing numbers of people, changing environments which influenced changes in eating habits, and possibly even magico-religious factors. On available evidence, Africa was by no means the first place where agriculture developed. The oldest evidence so far is only about 6400 years old, showing clearly that one form of wheat was cultivated in the Egyptian Fayum. Later, about 3200 BC, small domestic goats are known to have been kept by early farmers near Khartoum. People were beginning to concentrate into larger, more settled groups about this time, Merimde on the Nile Delta being an example of early village development.

The change to this new way of life probably spread first to the west, but seems soon to have extended into the Saharan area. These neolithic pastoralists, with cattle, goats, and sheep, added to their food supply by the older methods of hunting and fishing. The change to agricultural methods came slowly to other regions in Africa. It took about 2000 years for farming to extend successfully into West Africa and beyond. There is still much uncertainty about the antiquity of local plant cultivation in such areas, and some would argue that African types of rice were first used by the early farmers of the Middle Niger lake district even as long ago as 1500 BC. Perhaps most important in areas of rain forest was yam cultivation, for this plant thrives in such an environment whereas cereal crops do not.

A delicately modelled pottery head found at the early Nigerian site of Nok.

In more eastern parts of Africa changes were also taking place. During the first millenium BC, cattle- and sheep-owning peoples populated the Rift Valley area, and perhaps cultivated millet. Their society seems to have been a very successful one in Kenya and lasted until the 16th century AD, when iron-using Negro bands moved in and replaced them.

Movements of early Negroes towards the Kalahari were probably responsible for extending the range of cattle and sheep farther southwards. The great success of Bantu-speaking Africans – cultivators of root crops, millet, and sorghum – was another important factor in the cultural history of Africa. The origin of these people is unknown, although they possibly derived from western Africa. It was as a result of considerable Bantu expansions that Bushmen and Hottentot bands have been so severely reduced during the last thousand years.

The final important and widespread change to occur widely in Africa was the use of iron. This metal helped the development of some distinctive regional cultures in various parts of Africa. In northern Nigeria, for instance, the Negro people of Nok who lived between 500 BC

Left: This bronze figure, one of several, was found on Tada, a semi-island on the south bank of the middle reaches of the river Niger. According to the legends of the Nupe kingdom their founder, a mystical hero called Tsoede, rowed up the Niger in a canoe and left the figures on Tada where they became part of a cult of this ancestral hero. They are in a variety of styles of unknown origin except for this one, which is an Ife bronze of about 1300 AD probably cast by the cire perdue or lost wax method.

Right: A bronze head, dating from the 11th to 15th century AD, cast at Ife, in Nigeria. It may have been used in memorial ceremonies to a dead chief or king, mounted on a figure which would then be displayed.

and AD 200 were skilled iron users; they also had specialists in the community who produced beautiful terracotta art work. Such workmanship stimulated the later production of other works of great artistic merit such as the bronze heads from Ife. Quite clearly settlements were by then large, and there may have been quite complex tribal organization in some regions.

So the increasing use of metal, especially iron, provided the basis for a huge technological revolution. Metal assisted in many aspects of life: cutting timber, hunting, and warfare. In the later phases of African history some communities clearly evolved into cultures of some magnificence. The Zimbabwe culture in Rhodesia is one example. The people who moved to the site of Zimbabwe in about AD 300 were experts in building dry stone (mortarless) walls, as well as being skilled potters and gold miners. The region seems to have

Part of the ruins of the great Rhodesian city of Zimbabwe. The elliptical walling is somewhat unusual and shows the original thinking of its makers, although possible influences from the Arab world cannot be excluded.
Below left: A cast of a carved wooden bowl found at Zimbabwe.

been ruled by a monarch of considerable local importance.

We know almost nothing of the builders of Zimbabwe, whose civilization had vanished long before Europeans penetrated the African interior. This happened to any significant extent only from the 19th century on, although for hundreds of years European trading posts had been established on the coast. Unlike the first Spaniards to penetrate Central and South America, the explorers of Africa found no great city civilizations; the Africans had

Left: This carved palette was found at Hierakonpolis in Upper Egypt. The side shown here depicts King Narmer, wearing the White Crown of Upper Egypt, striking down an enemy; the group of hawk, head, and reeds is thought to signify that Narmer, the incarnation of the hawk-god Horus, takes captive the marsh-dwellers. To the left is the King's sandal-bearer and at the bottom two corpses. The reverse of the palette shows Narmer wearing the Red Crown of Lower Egypt, the corpses of rebels, mythical beasts, and the King as a bull destroying a town.

farmers of the fifth millenium BC gave way to more advanced *predynastic* cultures when village life developed, pottery became more refined, and the communities grew more artistically conscious as witnessed by their ivory spoons, bracelets, and combs, and the many beautiful vases they shaped from local stone. Many burials of these early Egyptians have been found, crouched and waiting for rebirth in a spirit world with pottery and other items beside them, presumably to help them on their journey into the

Right: A vase from the predynastic Gerzean culture (about 3300 BC). Many such vases have been found; the paintings of boats, people, standards, birds, and trees may illustrate episodes in religious ceremonies. The clay used is not Nile mud, but comes from desert valleys in Middle and Upper Egypt.

remained pastoralists, hunters, and warriors. But there was an outstanding exception to this – for one of the greatest of all the early civilizations grew up on the banks of the river Nile.

Egypt attains magnificence

The rise of an advanced civilization in Egypt is one of the most fascinating developments in the history of human societies. The varying stages in the growth towards city and state can be traced through excavations. The simple neolithic

This body, buried in predynastic Egypt, was dessicated and preserved by the hot, dry soil. Later attempts at preservation through mummification were seldom so successful.

next life. Religion and mythology were certainly playing an increasingly important part in their lives.

These predynastic cultures, the Badarian, Amratian, and Gerzean, are all named from sites in Upper Egypt – that is, south of Memphis. But how far each extended, and to what extent life differed in Lower Egypt (the northern part) is unknown. Probably there were independent communities each with its own god. These communities gradually formed larger and larger units through conquest, until there were only two kings: that of Upper Egypt, wearing the White Crown, and that of Lower Egypt who wore the Red Crown.

The Badarian and Amratian were purely African cultures; the Gerzean, for the first time, shows foreign influence.

135

Writing appears in this period, already in quite a sophisticated form; from about 3400 BC appear monumental buildings of mud brick in Sumerian style. Sumerian art motifs were used, and Sumerian cylinder seals have been found. There was evidently a great stimulus towards civilization from Mespotamia, but we do not know whether this was the result of colonization or simply of strong and frequent contact through trade.

These advances were made possible by the extraordinary fertility of the Nile Valley. Every year the river deposited quantities of rich black silt over its flood plain. When the waters had subsided, grain had only to be scattered and lightly trodden in to produce a tremendous yield. But irrigation projects had to be worked out to keep the crops watered; these demanded cooperation with and between communities. Large numbers of people began to give allegiance to a kingly individual and the early dynasties arose, in which the pharaoh was not only the ruler but something of a god as well.

The early dynasties

The division of Egyptian history into dynasties begins with the semi-legendary King Menes of Upper Egypt, who is said to have conquered Lower Egypt and so unified the two countries. The merging of the highly cultured life of the delta with the organization needed in the more harsh conditions of the south resulted in a society with a greater potential for civilization than had been known before. The unification took place in about 3200 BC; some scholars identify Menes with a king called Narmer, who left a hieroglyphic record of his victories on the palette of Hierokonpolis.

Angus McBride

Left: A slate hippopotamus, dating from the later predynastic period. Slate objects like this were often left in graves; they are frequently called palettes, but do not always show traces of paint or cosmetics. They may have had a magical significance.

During the Archaic period (Ist and IInd Dynasties) flexible paper or papyrus seems to have been invented, greatly easing communication; there is even evidence of works of reference from the period, a book on religion, and another on medicine. The complications of farming land that was annually flooded demanded accurate surveying, which encouraged the development of astronomy and mathematics. By the IIIrd Dynasty a system of weights and measures was in existence, and taxation had developed. But the history of these times is still shadowy.

Egyptian history takes on a more personal aspect with the IIIrd Dynasty, which, perhaps for the first time, sees the rise of a remarkable individual. This man

A predynastic village on the river Nile. Drawings on pots of the time show that the early Egyptians used papyrus reeds not only for their huts but also to build boats.

MUMMIFICATION

In early predynastic times, Egyptians were buried in pits dug in the hot, dry sand which preserved the bodies by natural dessication. Later, massive tombs of mudbrick and rubble were constructed for kings and nobles, but it was found that a body placed in such a subterranean burial chamber was far less well preserved than those in the older graves. As it was believed that the soul – or part of it – would re-enter the body, efforts had to be made to preserve it. In the Ist Dynasty bodies were wrapped in linen to try to maintain their form; by the IInd Dynasty, from about 2980 BC, the shape of the face and other parts of the body was retained by wrapping it in bandages soaked in some gummy substance. The limbs and fingers were not concealed in the binding but separately wrapped; there was no treatment of the body itself. The mummy was laid in the tomb curled up on one side, without the later elaborations of mummy-cases and head-dresses. The bandages dried quite hard and so the shape of the body was retained, but when these mummies are examined they contain only bones. Later processes were improved and elaborated.

The term 'mummy' comes from the Arabic word *mumia*, bitumen, since this substance was long thought to have been used in the process. In fact natron (sodium carbonate) was used.

Part of the Book of the Dead of the Priestess Anhai, written about 1100 BC to ensure the dead woman's safe arrival in the next world.

HIEROGLYPHS

The use of hieroglyphs began in Egypt in late predynastic times about 3400 BC. The earliest signs are found carved on religious palettes; as inscriptions on stones in tombs and buildings; engraved or painted on small labels of wood and ivory placed in tombs with objects for the dead; and as seal-impressions on pots containing food in tombs.

The origins of hieroglyphs are mysterious, for they first appear as a fully developed written language. A written language usually begins with pictograms portraying objects and simple actions, but this stage is missing in Egypt. Even the oldest inscriptions employ ideograms and phonograms (see page 110), and use signs to represent numbers. Although inscriptions showing an older stage may one day come to light, it is possible that the idea of writing using ideograms was introduced ready made from Iraq about 3400 BC, together with other influences for which there is evidence at this date.

From the time of the 1st Dynasty, about 3200 BC, hieroglyphs were painted on 'paper' made from two layers of split papyrus reed stems pasted together at right angles to one another. The signs were beautiful but intricate, so that writing even on papyrus scrolls must have been slow. Even in late predynastic times, a cursive script had been developed. This *hieratic* script was used for business, and, from the 7th century BC, a *demotic* (popular) script developed; this can be seen on the Rosetta Stone, a basalt slab of 196 BC which was founded in 1799 at Rosetta in the Nile Delta. On it is carved an honorific decree of Ptolemy V, repeated in hieroglyphs, demotic, and Greek. The French linguist Champollion realized the significance of the stone, and in 1882 deciphered hieroglyphs.

Two of the many treasures from the tomb of Tutankhamun: a little gold statue of a fisherman (above) and a gold mask of the young pharaoh inlaid with semiprecious stones. Many of the tombs of pharaohs were robbed not long after they were built, but that of Tutankhamun which was excavated only in the 1920s provide a vast treasure and has proved invaluable to archaeologists in reconstructing Egyptian civilization at its height.

was Imhotep, the vizier to Zoser, the second king in the dynasty; he lived about 2680 BC. In a way Imhotep is responsible for the modern idea of Ancient Egypt; he designed the first pyramid. The pyramids, apart from being the best known creations of this civilization, are part of a tradition of elaborate burial customs which is responsible for most of our knowledge of the life of Ancient Egypt. With the dead the Egyptians buried not only treasures like that of Tutankhamun and others that have been lost, but small models of everyday objects, paintings, furniture, and even mummified food. The pyramids themselves are evidence of a highly skilled and technical society, while the objects they contain have enabled us to build up a detailed picture of a truly advanced civilization.

The Americas

For well over 2500 years before the arrival of the Spanish conquistadores, a series of magnificent Indian civilizations flourished in Central and South America. Today, the jungles and highlands of this region are littered with the relics of these great cultures. The earliest empire to arise in Central America was that of the Olmecs. Their chief ceremonial centre was located at La Venta in Mexico, a humid, swamp-infested site on the Gulf Coast. Over the centuries, the damp climate and the jungle have worked to obliterate nearly every sign that men once lived here. This beautiful tile pavement bearing the design of a large jaguar mask has been preserved in superb condition by being covered over shortly after it was laid.

In AD 1519 a small band of Europeans landed on the coast of what is now Mexico They were the first of the *conquistadores*, the Spanish soldiers who, in a few years, conquered Central and South America and in doing so destroyed advanced civilizations of great complexity. In many ways these paralleled those of the Old World but they had evolved independently, since for many thousands of years the Americas had been quite cut off from the rest of the world.

Independent evolution of man in the Americas is ruled out by the absence of any fossil remains of primitive hominids or their primate forebears. And genetic research also shows that the native races of America are, like those of the rest of the world, of the species *Homo sapiens sapiens*

and share the blood groups A and O. It is thought that they are descended from limited numbers of advanced hunting peoples with Stone-Age technology, who crossed the land bridge which joined America to Asia up to 28,000 years ago.

Across the land bridge

There is still considerable controversy over the exact time and the nature of the initial colonization. There were two main phases during the Pleistocene when a land bridge between Asia and America could have existed at the same time as passes through the usually impenetrable glaciers in the mountain ranges of Alaska. Since remains of early people have been discovered dating from at least 25,000 years ago, the earlier phase of around 28,000 years before the present seems likelier than the later phase of 12,000 years ago.

Genetically these people are known as American Indian Mongoloid or Amerindian, and are distinct from the Eskimo who seem to be more recent arrivals (see page 76). The fact that present Amerindian groups entirely lack the blood group B and have very little A is an argument in favour of a single colonization by one family group, for since the peoples of Asia have all three groups, a colonization taking place over a long period with successive waves of immigrants would necessarily carry all categories with it.

Scant remains exist of these earliest peoples. There are sites in North America which show the passage of the migrants – ancient hearths with scatters of debris and a few tools that point to an economy based on the hunting of large game such as elephant, bison, elk, and horse. Carbon 14 dates of these sites show, as might be expected, that there is a trend of earlier to later from north to south. Old Crow Flats and Onion Portage in north Alaska date from about 24,000 and 11,000 BC respectively. At Tlapacoya in Mexico a date of 22,000–20,000 BC is given to flake tools, showing the early passage of man down through Central America on his way south. By 10,000 years ago, these hunting peoples had ventured right down to the tip of South America.

Throughout the Ice Age, or Pleistocene Period, the New World inhabitants followed a pattern of life which archaeology traces through their discarded tools and weapons and in the remains of the animals they slaughtered. At Folsom in North America, flint points were found associated with an extinct species of bison; Dent in Colorado and Santa Isabel Iztapan in the Valley of Mexico are both mammoth-kill sites with butchering implements and projectile points of flints and obsidian lying in and among the animal bones. And archaeologists have been able to work out the exact time of year and the wind direction on the day primitive hunters stampeded a herd of bison into a ravine at Olsen Chubbock; the remains of nearly 200 animals were found piled into the narrow ravine where they died and were butchered. Cave sites in the Ayacucho Valley in South America have similar traces of hunters who preyed on a species of giant ground sloth, now extinct.

Settled agriculture

By the Neothermal – 'New Warm' – around 12,000 years ago, these early hun-

The first inhabitants of America wandered on to the continent from Asia across the land bridge that spanned the Bering Strait. Over the centuries they drifted southward to occupy the entire New World, and the routes along which they spread can be traced by the remains of their ancient camp sites. Bits of bone, odd scraping and cutting tools, and a variety of arrowheads and spearheads have been unearthed from these temporary dwelling sites. Archaeologists, piecing together the clues, have been able to reconstruct a great deal of their life. Their Stone Age tools and weapons show these people to have been highly skilled hunters. Fluted flint spearheads or 'Clovis points' (above) were shaped by striking pieces of flint with rock hammers or by using bits of pointed bone to chip off a great number of very small flakes. In this way, a sharpened point and regular bevelled edge could be obtained. Such flint weapons were especially common on the plains to the east of the Rocky Mountains. Here the Indians hunted a wide range of big game that included mammoths, mastodons, and bison.

The original wild form of maize is no longer known to man and the story of its early cultivation has been lost, but these ancient cobs from the Teohuacan valley Mexico show stages in its development. Surprisingly, maize began to be raised rather late compared to such ancient crops as avocados and chili peppers. The size of an ear of maize was at first no more than a few centimetres in length. With careful breeding over the ages, the plant has increased enormously in size.

ters had undoubtedly evolved a sophisticated technology based on finely flaked projectile points of stone. But their way of life was necessarily a mobile one, pursuing the large plains animals and gathering wild plant fodder. The evolution of agriculture in the New World, as in the Old, was by no means a simple process, but gradually man's economy shifted from one based on the pursuit of large game animals to one of dependence on three major crops which supported the later civilizations – maize, beans, and squashes.

The change in climate which occurred about this time may well have been an important factor in initiating this slow but revolutionary change, for the Neothermal saw the extinction of some of the major large plains animals. The horse disappeared and was only reintroduced several thousand years later by the Europeans; the elephant died out altogether; and the bison retreated until its southernmost area was the Great Plains of North America. As North Americans adapted themselves to the changing environmental conditions several different cultural traditions emerged.

The so-called *Big Game Hunting Tradition* represents the continuation of the former specialized hunting economy, now restricted to the North American Plains. In the Pacific North-west, extending down as far as California, the *Old Cordilleran Tradition* shows how men adapted to the rich off-shore marine life, supplemented by the fishing of inland rivers, the trapping of game, and collection of nuts and acorns in the forests. Neither of these saw the development of agriculture, partly because the environment was not always suited to the raising of maize and not least because in both cases the Indians gained a more than adequate living from their hunting and gathering so there was little incentive for them to take up new ways of acquiring food. The *Archaic* and *Desert*

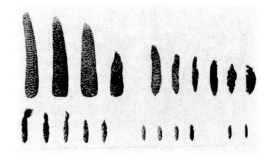

Traditions refer respectively to the hunting, fishing, and collecting economy of the Eastern Woodlands, adapted to conditions of temperate forest and river environment; and to the pattern of subsistence activities in the arid West, especially the Great Basin, whose inhabitants supplemented the collection of wild plants and seeds with the pursuit and trapping of smaller game. The Desert Tradition in particular saw the gradual shift in some areas to agriculture, especially in the American South-west and Mesoamerica.

In the Tehuacan Valley in Mexico, research has yielded a great deal of important information about the evolution of food production. Early inhabitants of the valley appear to have followed strictly seasonal routines; they moved within roughly territorial boundaries in extended family units – macrobands – during the late spring, summer, and autumn, living off leaves, pods, young shoots, and fruits. During the lean and dry winter season, they would split again into single family groups – microbands – to hunt deer and rabbits with spear-throwers. Remains of the camps they left behind them give archaeologists clues to their habits – the number of hearths in a camp suggests the size of the group, with the food remains, in many cases well preserved in the dessicated conditions, showing the time of year at which the camp was occupied.

The change from gathering wild plant foods to harvesting planted crops was a very gradual one. At first many species were exploited – the bottle gourd; the chili pepper; the avocado; various squashes and beans; grasses such as mesquite, teosinte, and amaranth; and the fleshy leaves and flowers of the maguey cactus. These plants continued to be grown but there was an increasing concentration on a few of them – the ones that underwent genetic changes rendering them many times more productive. The most important was maize.

As food production became more efficient through the cultivation of a few highly productive plants and the advent of irrigation systems, the population rose; this is shown by the increase in the number and size of archaeological sites in the Tehuacan Valley. People ranged less far afield in a year, and concentrated their activities in one or two main regions. As they learned how to produce crops plenti-

The Andes are a vast store-house of mineral wealth; seams of gold, silver, copper, and tin abound in the region. The Indians of the Andean highlands were skilled metal workers from very ancient times. They turned out great numbers of beautiful figurines such as this gold llama. For hundreds of years gold was only worked by beating or hammering it into stone moulds. It was not until the 4th century BC that smelting became a commonplace technique.

This gold disc portrays the head of the Earth Goddess surrounded by a wheeling cluster of maize, yucca and sweet potatoes. The various sections of the disc form the divisions of a sowing calendar for these crops. Dating from the 12th or 13th century AD, it was made by the Chimu of northern Peru.

ful enough to last them through the lean months of the winter, these groups of people became more settled. We can trace the development of the first hamlets and villages in this region around 1500 BC, while semi-permanent communities probably existed before 3000 BC.

The Tehuacan Valley has been used as a stereotype for the evolution of farming and a settled way of life for other parts of the Americas. Although this is sometimes valid, it is dangerous to simplify a highly complex process such as this to make it universally applicable. On the coast of Peru for example, settled village life evolved earlier than food production because of the plentiful supplies of fish and other marine creatures. Large populations were present in relatively settled communities as far back as 2500 BC and possibly even earlier. The coast communities appear to have turned increasingly to agriculture, perhaps as the food from the sea diminished through over-exploitation; but crop plants including the gourd, squashes, lima and jack beans, amaranth, cotton, maize, and high-altitude plants – the potato, the sweet potato, and the cold-climate grain quinoa – were nearly all first cultivated in the Andes and only later introduced to the coast. The llama and alpaca, guinea pig and muscovy duck were similarly domesticated in high mountain

The Anasazi Indians of the American South-west built their homes in inaccessible perches to protect themselves from Apache and Navaho attack. Mesa Verde in Colorado is one of the most spectacular examples of a Pueblo village. This cluster of adobe and stone houses was virtually invulnerable to attack from either above or below.

The main cultural traditions that had developed in North America by 1500 AD. 1 Arctic Eskimo hunters. 2. Sub-arctic hunters. 3 North-west coast fishers and hunters. 4 Interior plateau river fishers and hunter-gatherers. 5 Great Plains bison hunters and semi-permanent farmers. 6 Woodlands temperate hunter-gatherers and farming groups. 7 Mississippi and Ohio valleys permanent farming communities. 8 California coast and valley hunting, fishing, and collecting groups. 9 Great Basin desert gatherers. 10 South-west sedentary farmers and gathering groups. 11 Desert gathering groups. 12 Mesoamerican urban states.

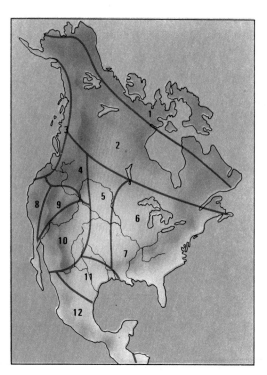

regions; they represent the only domesticated animals in Pre-Conquest South America beside the dog.

North America

The North American Indians are distributed throughout the whole continent, from the Arctic circle to the Gulf of Mexico, excepting only the most extreme of desert conditions. The diversity of cultural traditions in prehistoric North America bears witness to the amazing way that man can adapt to different environments, including Arctic tundra, forest, mountain, plains, and desert. The well-known bison-hunting

Plains Indian is only one such adaptation; and his culture is only as recent as the introduction of the horse and gun with the first European settlers early in the 17th century. All these traditions can be traced back to the four main types which, as we have seen, emerged between 10,000 and 5000 BC.

Between that period and historic times, the Indian cultures evolved continuously. Some groups responded to impulses channelled up from Mesoamerica; these took up cultivation, pottery-making, and the building of large settlements with fortifications and ceremonial buildings. This civilizing influence can best be seen in the Anasazi and Hohokam cultures in the South-west between the early centuries AD and the 14th. The Hohokam built long irrigation canals into the deserts of central Arizona. The Woodlands culture of eastern North America also saw adoption of agriculture and the development of sedentary and semi-sedentary communities. Here the Adena and Hopewell cultures (centred about the Ohio basin) spread their influence widely, building large burial mounds over the whole area as far as the eastern edges of the Great Plains. The Hopewell excelled in hammered copper work – probably the finest metalwork produced in North America.

Around AD 700 Hopewell influence declined, as the Mississippi culture spread throughout the area centred on the Mississippi, Ohio, and Missouri rivers. The cultivation of maize was especially important to the economy of this tradition; it resulted in the development of larger permanent communities than those of the Hopewell or Adena. When Europeans first entered the Eastern Woodlands area, the Mississippian culture dominated much of it, merging on the northern periphery with the Woodland tradition (temperate forest collecting and hunting groups) and to the west with the Plains economies.

The great North American Plains saw a similar steady cultural evolution from the Big Game Hunting and Archaic traditions; in some parts cultivation was adopted and in others, like the prairie lands to the west, earlier hunting patterns continued. Life on the Plains was to be revolutionized by the arrival of the horse.

On the Pacific Coast farther west, hunting, fishing, and collecting groups were little influenced by the rise of agriculture elsewhere. The acorn gatherers of the California woodlands maintained quite dense populations in relatively stable settlements, while the North-west Coast and Plateau traditions, which had emerged by the first millenium BC, saw flourishing communities around the sea-shore and rivers with their plentiful supply of fish. Wood – pine, fir, and redwood – was plentiful, and a distinctive style of carving developed, especially seen in the intricate totem poles. On a more practical level, sea-going canoes were built, some up to 18 metres (60 feet) long. Contact with the first European settlers gave rise to even greater material wealth as the fur trade developed. These communities were more rigidly organized than most Indian groups; offices like that of chief were open only to members of certain families, rather than being filled on merit. The aristocratic nature of society on the North-west Coast is seen in the *pot-latch*, a celebration given by high-ranking families where they would give away or even destroy much of their wealth.

Farther north, in the vast stretches of sub-Arctic coniferous forest and Arctic tundra and coast, Indians and Eskimo-Aleuts hunted caribou, elk, seal, walrus, and sometimes whale, with chipped flint and bone tools. The true Eskimo tradition had emerged by 1000 BC. These people moved about easily with the aid of canoes, and when river travel was impractical, made use of dog-sleds and snow-shoes. The forest zone supported the rich Denetasiro culture, based on hunting, fishing, and the trapping of small animals.

Certain artefact types are common to many North American Indian groups. Pottery was fairly widespread, although neither the potter's wheel nor glazing was known. Chipped and finely flaked stone tools such as projectiles, knives, and axes were common by the first centuries AD. Textiles were woven from cotton or other plant fibres.

Village communities

Whether settled or nomadic, the Indians sometimes lived in fairly large communities – the villages of sedentary peoples or bands of hunters or nomads. There were many different types of dwelling, depending on the habits and environment of their owners, but a few types recur. The conical bark or skin *tepee* was widely used by nomads like the Plains hunters and some of the Indians of the Far North. Its long poles could easily be carried on a canoe, or, as on the Plains, form the base of a dog-sled. But most Indians, being at least semi-sedentary, preferred more solid homes. The most common were the *wigwam* type, where a dome- or tunnel-shaped frame was covered with skins, bark, or brush; and earth lodges, built in pits and roofed with turf. Both could be easily built if the community moved. The most permanent houses built by the North American Indians were the plank houses of the North-western traders, and the *adobe* (mudbrick) villages of the Pueblo farmers in the South-west. The latter consisted of structures like large groups of flats with as many as several hundred rooms, rectangular or semi-circular in shape and surrounding a central compound area. They were entered by holes in the roofs, reached by ladders which could be pulled up in case of attack.

Horses, introduced by European settlers, spread rapidly through North America, and by the 18th century mounted Indians were often seen.

The Indians of the north-west Pacific Coast were renowned for their ornate wooden carvings. The totem poles that stood outside their wooden dwellings were covered with grotesque animal figures – symbols of the tribe or clan to which each household belonged. Above is a painted thunderbird of the Kwakiutl tribe.

The Iroquois 'Nation' of the north-eastern woodlands was a federation of five Indian tribes. The Iroquois lived in permanent villages, frequently surrounded by a palisade of wooden stakes that gave protection from raiding enemies. On small plots of land surrounding the village they raised beans, squash, and corn (maize). Their diet also included a regular supply of game and fish. The Iroquois built long, rectangular dwellings called 'longhouses'. Inside each lived several families, each with its own cooking hearth and living area. This drawing by the Englishman John White, made around 1585, shows several scenes from the daily life of the Iroquois village of Secoton.

Undoubtedly the most affected by their introduction were the Plains Indians; their efficiency as hunters was greatly increased and many tribes turned from a mixed farming and hunting economy to the exclusive pursuit of the great bison herds. Tribal organization varied during the year as the herds came together and dispersed. The Plains hunters were efficient but economical; they did not kill more bison than were needed and made use of every part of the animal for food, clothing, tents, and tools and weapons. This pattern was shattered by the coming of the railways in the mid-19th century; the herds were systematically slaughtered by white men, at first to feed the gangs building the railways and later for sport. By 1889 there were 540 bison left and the Plains economies were totally disrupted.

The near extinction of the bison is the most extreme example of the destructive effect of European contact on the Indians of North America. But already by the 19th century many of these diverse peoples had succumbed to its impact. Old tribal territories, subsistence patterns, and cultural identities were disrupted until today very few native Americans exist. The survivors, for the most part, live in the sub-Arctic zones or on reservations, poor remnants of the vast territories through which they once roamed.

Long before the rise of the ancient Greeks, the Olmecs had already established a great civilization around the Gulf Coast region of Mexico. They were master stonemasons who produced some highly distinctive pieces of sculpture. A number of massive stone heads with flattened noses and thick lips has been recovered in recent years. These giant statues weigh up to 20 tonnes and stand over 3 metres high. The heads, clad in curious helmets, are thought to portray warriors or ballplayers.

The wheel did not figure in the technology of the New World before the coming of Europeans – probably because the only potential draft animal, the llama, lived in steeply mountainous regions. That the wheel was not unknown is proved by this little Aztec toy.

Complex civilizations

The two main regions in the New World which saw the growth of advanced civilizations are Mesoamerica (or Central America) and the Central Andean and neighbouring coast of South America.

The empires which grew up here, those of the Aztecs, Maya, and Inca, are the end products of two related and similar cultural traditions which had given rise to other earlier rich and sophisticated cultures. These different civilizations had a number of points in common including a level of technology which had given them fine pottery by around 1800 BC (earlier in some quarters), and a shrewd knowledge of metallurgy, although never developing beyond a Bronze-Age level. Monumental architecture in mudbrick and stone developed early and in both areas the concept of pyramid and platform mounds planned around open courts and plazas is a fundamental one. Although the true arch was unknown, architectural skill was considerable, as shown in the fine temples and pyramids of the Maya and the builders of Teotihuacan in Mexico, and in the superb buildings of the Incas in Peru, the walls of which, resembling an intricate jigsaw in stone, are able to withstand the numerous and often catastrophic earthquakes in these areas.

The Mexicans and the Maya both had pictographic writing systems with which they recorded political, economic, and especially astronomical events. The surviving documents are in the form of *codices*, in which the hieroglyphic symbols are painted on skin or bark cloth. The South Americans had no such recording device, yet the Incas and possibly earlier peoples used the *quipu* – a mnemonic system of knotted strings of different lengths and colours by which they kept their records and accounts.

The wheel plays no part in the technology of the American civilizations. This does not mean that they did not know about the wheel – a child's pull-along toy from Mexico shows that they did. But the Central and South Americas provide no large animals to domesticate as beasts of burden; and the nearest – the llama – is unsuited to any work but carrying light loads and lives in high mountainous regions where wheeled vehicles would be of little or no use.

More than anything, American civilizations share certain social and political attributes which, while evolved without contact with the Old World, nevertheless make the nature of the New World developments recognizable. Division of labour, developed class structure, different religious and political concepts, and institutions of royal, priestly, legal, or military natures are all New World independent parallels to what we see as Old World precedents.

Ancient Mexico

Between the evolution of agriculture with settled village life and the rise of Aztec suzerainty over much of Mesoamerica, there existed a medley of successive civilizations which are divided into different chronological phases. The first recognizable civilization to emerge is that of the Olmec, in the swampy tropical forest regions of the Gulf Coast. C_{14} dates from two of their major centres at San Lorenzo and La Venta show that these people flourished between 1300 BC and 400 BC spanning the Formative or Preclassic Period. This era sees the crystallization of the Mesoamerican cultural tradition with the first large-scale building of pyramids and plazas, temples and ball-courts, and the development of the sacred game played there. Art styles and religious themes emphasized the strong and the fearsome with jaguars, eagles, and snakes and imaginary creatures all symbolizing these qualities. All these elements are present from Olmec times onwards.

Olmec was not an urban culture but like the Maya was based on ceremonial centres which acted as political and religious focuses for scattered villages whose inhabitants practised a shifting pattern of cultivation called *swidden*. Olmec art is characterized by a roundness and heaviness with many fine ceramic and stone figurines of babies or with baby-like characteristics. Monumental stone representations of human heads with curious helmets, possibly ball-players, are peculiar to the Olmec tradition.

The Olmec culture declined and fell to outside pressure with first San Lorenzo and then La Venta being abandoned. In other parts of Mesoamerica contemporary Formative cultures, often linked with the Olmec through trade-routes, underwent

THE MESOAMERICAN BALL GAME

The ball game was popular throughout Mesoamerica, probably from Preclassic times until the Conquest. Ball courts were important features of every large city or ceremonial centre and the game which Spanish chroniclers describe from Aztec contexts evidently had a religious as well as sporting significance.

The court was shaped like a capital I and the purpose of the game was to propel a solid rubber ball, from 15 centimetres (6 inches) to nearly 30 centimetres (1 foot) in diameter, which could weigh as much as 2½ kilogrammes (5 pounds), across a line drawn down the centre. Players were usually professionals and were drawn from the nobility. They wore special cotton padding about their heads, waists, and joints, since the game allowed the use of the body alone for striking the ball, and injuries were frequent. Should any player succeed in driving the ball through one of the stone rings set high up at either end of the court, his team won the game outright, whereupon they had the right to claim the clothes and jewels worn by the spectators – if they could catch them in the mad stampede for the exit that inevitably followed.

The game was often played for purposes of divination. It is recorded that Montezuma II played against the Lord of Texcoco to determine whether his opponent's astrologers were correct in predicting the subjugation of Mexico by foreigners. He lost; and as history bears witness, he was later deposed and executed by the Spaniards who subsequently ruled his kingdom.

The Toltecs were a fierce warrior tribe that swept down from northern Mexico around 1000 AD and rapidly overran the Olmecs. These warlike people were themselves soon weakened by an endless round of internal struggles and wars. Their capital at Tula was eventually destroyed in 1168 AD. Throughout the Toltec era, an enormous variety of metal ornaments and beautifully worked jewellery was produced. This double-headed Toltec serpent is typical of the fierce mythical beasts that fired the imaginations of the people of the time. The serpent's body is covered with a mosaic of turquoise pieces intended to resemble the scales of a real snake.

This map shows the location of the main civilizations of Mesoamerica.

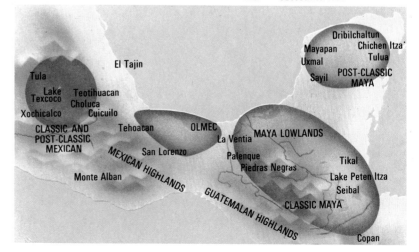

	MESOAMERICA		SOUTH AMERICA	
	MEXICO	LOWLANDS	HIGHLANDS	COAST
AD	CONQUEST			
1500	Aztec Empire		Inca Empire	
	More invasions	Mayapan	Various	Chimu Empire in north
	Toltecs at Tula	Chichen Itza	Late Intermediate	
1000	Chichimecs	Collapse of Classic centres	Huari and Tiahuanaco Empires	Pachacamac
	Post-classic		Middle Horizon	
	Collapse of Teotihuacan	Ceremonial centres eg Tikal, Copan, Palenque		Mochica in north
500	Expansion of Teotihuacan			Nazca in south
	Rise of Teotihuacan	Classic		Valley irrigation works
0	Cuicuilco		Early Intermediate	
BC				
	La Venta	Rise of ceremonial centre tradition	Chavin 'Cult of the Cat'	Paracas Necropolis
1000	Gulf Coast Olmec			Larger ceremonial centres
	San Lorenzo	Formative	Early Horizon	
2000	Permanent villages	First pottery		Small-scale irrigation
				First pottery
	First pottery			
3000	Semi-sedentary communities		Pottery in Ecuador	Large permanent villages
	Beginnings of agriculture			
5000	Hunting and gathering			Early fishing communities

Below: A number of massive carved columns in the shape of warriors and feathered serpents upheld the roof of the Temple of Quetzalcoatl in the Toltec capital of Tula. Some of these ponderous, 5-metre figures were restored during the exploration of the site.

Below right: The ruins of the once great city of Teotihuacan form a vast complex of temples and buildings.

similar changes giving rise to the Classic Epoch around AD 300. During five centuries leading up to this time the greater part of Mesoamerica and especially Mexico was made up of small but growing city-states jostling for power among themselves. Of these, Cuicuilco near Lake Texcoco was a powerful centre. Not far away was Teotihuacan which was to become one of the mightiest in all Mesoamerica during the Classic period.

The critical point in the evolution of a city is when it becomes too large to be supported by the surrounding area. This happened to Teotihuacan by about AD 500 when the population is estimated at more than 50,000. From then on the archaeological record shows a marked increase of foreign affairs and the rapid and sometimes possibly violent spread of Highland Mexican influence throughout the whole of Middle America.

Teotihuacan was a city in the true meaning of the word, having not only a huge ceremonial quarter but also a large urban population, all grouped within the 21 square kilometres (8 square miles) of the city's boundaries. Building within this area was planned to a central grid and all the classic qualifications and functions of a city were present – an elaborate and highly stratified class structure with secular lords and priest dignitaries at the head of a large and well-organized bureaucracy, a standing army, and an artisan class organized within different quarters of the town according to their trade. Other great cities contemporary with Teotihuacan, representing rich and powerful cultures outside Central Mexico, were El Tajin which superseded the Olmec on the Gulf Coast and Monte Alban in Oaxaca on the Pacific side of Mesoamerica.

The Post-Classic period was ushered in by the fall of Teotihuacan, possibly overthrown by the war-mongering Chichimec peoples who harried the borders of the civilized world from the northern marches of Mexico. Like many barbarians responsible for the collapse of existing civilizations, these Toltec warriors later settled down; in this case, the Toltecs built their own city not far from the ruins of Teotihuacan at a place called Tula. The Classic period may not have been particularly peaceful, but the Post-Classic era was definitely pervaded by a spirit of militarism and a ferocity of religious belief which lasted from its outset until the height of Aztec domination. Grim *tzompantli* – skull racks denoting the cult of the severed head – make a début in the engraved wall reliefs. For the first time human sacrifice has importance in the rituals of society.

Brief supremacy

The Aztecs began rather ignominiously for a tribe of people later to hold most of Mesoamerica under its tyrannical sway. Their great capital of Tenochtitlan was founded in a swampy islet in the middle of Lake Texcoco, to which they had been driven by other Chichimec tribes when they arrived in the 12th century in a final wave of invasions. The Aztecs eventually dominated rather than administered an empire of many tribes who retained their freedom at the expense of a ruthless tribute to their

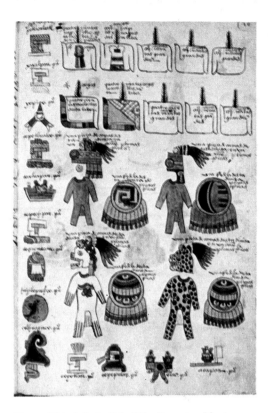

Above: The Aztec 'Calendar Round', which was invented in the 6th century BC, was based on a 52-year cycle. Each year of the cycle was divided into 365 days, or 18 months of 20 days each. The left-over period was a time of five 'unlucky days'. The signs of the 20 days of the month can be seen in this stone calendar, in the inner circle of rectangles.
Below left: This Mixtec obsidian knife has an exquisitely inlaid handle in the shape of a kneeling man. Ceremonial knives such as this were used for the human sacrifices that were carried out by the high priests of the Sun God Huitzilopochtli. The Sun God required a constant supply of human blood to carry out his duties to man and nature.
Bottom left: Skilled Mixtec craftsmen produced many of the ornaments used by the Aztecs. This shield was made out of wood with a mosaic of brilliant feathers.

Above: This tribute list of the Emperor Montezuma details quantities and types of goods and the areas they came from, giving an insight into the economy and the often perishable resources – cacao, maize, jaguar skins, and feathers – which generally leave no traces for archaeologists. The writing is a translation added by the Spaniards.
Below: This human skull overlaid with turquoise mosaic was part of the Aztec treasure collected by Cortes when he invaded Mexico.

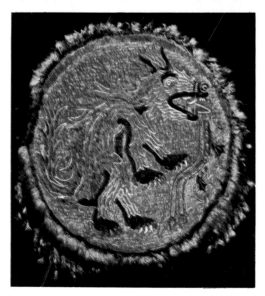

The ancient Maya site of Tikal in Guatemala was the greatest of the four major towns of the classical period of Mayan civilization. As early as the 1st century AD, Tikal had become a major ceremonial centre. The earliest Maya stele to date has been found in there. The inscriptions on it correspond to the date 292 AD.

The earliest indications of the practice of sacrificing human beings to the gods appear in the Toltec capital of Tula. These rituals became increasingly elaborate during the last centuries of the Toltec and Aztec empires. The mounting of severed heads as trophies, on skull racks called tzompantli, was also a practice of this period. Wall carvings that depict these racks have been found throughout the Mexican region and spread as far south as the Mayan town of Chichén Itza.

conquerors – a wealth of luxury goods, foodstuffs, raw materials, and human beings for sacrifice to Aztec gods.

Aztec supremacy was to be short lived. They had emerged supreme in a triple alliance with two other Texcocan cities in 1428, after a long struggle for survival and ascendancy. But 1519 saw the arrival of 'fair bearded warriors from across the seas' – a legend long established in Mesoamerica – and in a short time native American civilization had been annihilated by these Spanish *conquistadores*. Smallpox, as always one of the most potent weapons of the European invaders, reduced the population by as much as 90 per cent and more within 30 years of the Conquest. Demoralized and debilitated, the Mesoamericans were overwhelmed by the new culture and its religion.

The Maya

The civilization of the Maya people, which grew up in the dense lowland tropical rainforests of Guatemala and the Yucatan

Peninsula, remained rather distinct from developments elsewhere in Mesoamerica. Archaeologists have long questioned the emergence of such a sophisticated culture in so seemingly inhospitable a region. If great civilizations depend on intensive agricultural systems, how did the Maya manage to support themselves in an environment which appears to rule out any but the extensive and shifting pattern of cultivation adapted to such conditions? In fact, it is just beginning to be appreciated that this system of long-fallow farming is more productive than had been thought, and present estimates suggest that one man's efforts would yield enough to support up to 12 craftsmen.

The Maya civilization is typical of the kind which produced elaborate ceremonial centres, but little city development. Tikal, one of the largest Maya centres known, was sited in the densest of tropical forested regions and had an estimated maximum population of 45,000. At the same period in Mexico, Teotihuacan had a population at least twice as large. The distribution of these sites reflects a highly complex and finely balanced social structure. While Mexico had populations concentrated

within urban city states, the Maya adapted to their very different environment with 'extended states' with ultimately the same function. The Maya certainly had a great and prosperous civilization. They built superbly tall pyramids, temples, palaces, and ball-courts all decorated with fine and elaborate stone sculpture. Their unique and beautiful art style flourished in the working of jade, serpentine, and turquoise; in painted decoration of pottery vessels and in the production of figurines; and in the engraving of stelae, the stone monoliths found everywhere in Maya ceremonial centres. They were adept in the mysteries of astronomy, having early on calculated the precise number and fraction of days in the year; they built observatories and had simple instruments with which they calculated the Venus cycle and even, it appears, could predict eclipses of the Sun. They used a form of hieroglyphic writing and a system of 'long-count' which enables scholars to date the many events in Maya history quite accurately. This method of counting encompassed a basic principle not even known to the Romans – the concept of zero.

Maya civilization can be divided into similar periods to that of the rest of Mesoamerica. During the Formative and Classic periods, between 1000 BC and AD 900, the tradition of ceremonial centre building was established within the low-lying forests. The close of the Classic sees the inexplicable collapse of civilization in the major ceremonial centres. Great centres like Tikal and Palenque were depopulated and no more stelae were erected. The Post-Classic period saw political leadership pass to the north of the region with the rise of Chichen Itza. The most notable event of this period is the sudden infiltration of Mexican influence, recognizable in different architectural styles and in such practices alien to the Classic Maya as human sacrifice and the worship of the Mexican plumed serpent Quetzalcoatl.

When the *conquistadores* arrived early in the 16th century they found Maya civilization in an advanced state of decline; the old social and political order had long since disappeared. The political scene on the eve of the Conquest was marked by the presence of small individual city-states waging petty warfare among themselves. Small towns mark the late and somewhat ill-adapted development of urbanism in this part of the world; many are fortified, indicating troubled times, with the Mexican principles of extortion and tribute quite widespread. These regimes could offer little resistance to the Conquest.

The southern civilization

The first great civilizations of South America grew up in what is now Peru. Although this area lies only just south of the Equator, the heat is tempered by the Andes mountains which rise steeply within a short distance of the coast and run parallel to it, and by the cold Humboldt current which flows northwards along the coasts. Because of this current Peruvian coastal waters are extraordinarily rich in marine life, and settled fishing communities developed here early on; the remains of simple platform mounds of mudbrick have been found dating from around 2500 BC.

Simple agriculture is possible in the flood plains of the many rivers which dissect the coastal strip at regular intervals, flowing west from the Andes to the Pacific. After the development of irrigation about 1800 BC, the first really large settlements appeared on the coast and in the valleys. Terracing of the steep valley sides higher in the mountains was mastered early on and developed into vast systems such as those around Pisac.

Above: The palace at Palenque consisted of two courts bounded by rows of chambers. Its tower is 12 metres high. Palenque was one of the great Maya centres of the Classic Period.

The Peruvians rapidly mastered the terracing of steep mountainsides on which they grew their crops.

149

Above: The Peruvians were outstanding goldsmiths. This puma dates from the 4th to 9th centuries AD; the decorations on its body represent the sacred two-headed serpent.

Below: The Mochica culture of the northern Peruvian coast was renowned for its pottery, particularly portrait vases like these, depicting important lords and priests.

The introduction of pottery around 1800 BC in Peru heralds the first major phase of cultural development, called the Initial Period. The following Early Horizon saw the all-pervading influence of the Chavin art style, unifying in an extraordinary way the different cultural groups then present in Peru. Chavin was a highland and largely religious phenomenon, sometimes called the Cult of the Cat because of its snarling fanged imagery. It subtly pervaded many later cultures and concepts right down to the Inca period. The term 'Horizon' signifies a time when one particular culture, through religious or military influence, cuts across regional distinctions as a unifying force; intervening phases are known as 'Intermediates'.

During the First Intermediate, two areas predominated – the north part of the coast peopled by the Mochica and the south coast with the Nazca culture. In both, pottery manufacture had already reached a very high level; many different vessel shapes were produced, often from moulds (the potter's wheel was unknown), and decorated in highly burnished polychrome fired on to the body. More than anything, these Indians were renowned for the great quantity of fine textiles they wove which the arid conditions of the coastal desert have helped to preserve.

Between AD 600 and 1000, the Middle Horizon Period saw the rise of the great cities of Huari and Tiahuanaco in the Central and Southern Highlands. The primarily religious influence of Tiahuanaco on the shores of Lake Titicaca was taken up by Huari and spread widely throughout the highlands and along the coast, probably through military intervention. The pre-eminent deity was the 'Gateway' god from Tiahuanaco with his winged attendants, and these appeared everywhere. Even after the decline and fall of the two cities and the break-up of the empire, they continued to influence the development of the regional cultures although in a very derived and debased form.

It was with the Middle Horizon empires that true urbanism came to the coast, but this was a short-lived phenomenon in the south. In the north, the succeeding Late Intermediate Period saw the rise of an empire second only to that of the Incas who later conquered and subjugated them – the Chimu. Their capital was Chanchan in the Moche Valley; with its ten great walled compounds built of mudbrick and the surrounding network of urban building, fields, and irrigation systems it is one of the largest Pre-Conquest South American cities. It is to the Chimu that the Incas owed their command of metallurgy, for these people worked huge quantities of gold and silver and bronze. They knew the techniques of lost-wax casting, alloying, soldering, and repoussé. They were responsible for major irrigation works linking several neighbouring valleys. Their black burnished pottery was mass produced in a variety of shapes and styles, reminiscent of the earlier Mochica.

The Incas, like the Aztecs of Mexico, started as just one of a number of small warring tribes in the South-Central Highlands. After beating their neighbours, the Chanca, at Cuzco in 1438, they embarked on a programme of conquest still in progress at the Conquest. Their empire stretched for more than 3200 kilometres (2000 miles) – from the northern borders of modern Ecuador down into Chile in length, in breadth spanning the Andes from their east side down to the coast on the west. At its maximum, the Incas must have ruled over six million people.

Of all the Peruvian civilizations that of the Incas arouses the greatest interest, admiration, and respect. Their language, Quechua, was official throughout the empire and universally spoken. Their system of government was typically pyramidical – a theocracy ruled by a god-king who

traced his descent in unbroken line back to the first Inca, Manco Capac, and through him to the Sun. The Inca army was as powerful and as efficiently organized as the regime responsible for its creation. Two great highways ran the entire length of the empire, one on the coast and another in the mountains; along them at set intervals were wayside rest stations for the refreshment of moving troops on the march and runners who carried messages from one end of the empire to the other. The only area to remain invulnerable to Inca embassy or warfare was the tropical forests of the Amazon Basin to the east of the Andes. The last emperor, Atahuallpa, had emerged successful after a five-year civil war when he received the news that strangers had penetrated his kingdom. Within a few weeks he was defeated and held captive at the hands of the Spaniard Francisco Pizarro, and later he was executed. Over the next few generations there were several desperate but mainly unsuccessful attempts by the Peruvians to rid themselves of the foreign regime. European diseases such as smallpox took a hideous toll. And much that was unique and magnificent in Pre-Conquest South American civilization suffocated beneath the cloak of European culture.

Above: A decorated gourd, inlaid with turquoise and mother of pearl, which dates from the 12th or 13th century AD. It was made by the Chimu, who reached the coast of Peru from the north during the 11th century and built up a great empire. They were responsible for many irrigation works and were skilled metal workers; many of the feats for which the Incas are remembered were based on Chimu achievements.

A fragment of openwork cotton textile, showing stylized cats and fish, probably made by the Chimu.

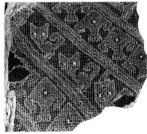

Left: The Inca were superb builders; great blocks of stone were shaped and ground away and fitted together without mortar – but so closely that a knife-blade could not be slid between them. Despite the many earthquakes in the region, many of their buildings still stand today.

151

Glossary

ACHEULIAN A term used especially with reference to certain earlier stages of the palaeolithic (Old Stone Age) in Europe and Africa. It is characterized by the occurrence of well-shaped stone handaxes. The site which gave its name to this cultural phase is Saint Acheul, Amiens, in France.

ALTAMIRA A cave in north-east Spain, south of Santander, which has some of the finest examples of palaeolithic art so far known. The animals portrayed include deer, bison, and wild boar.

ARAGO (Tautavel) A cave site in the French Pyrenees where excavations have produced artefacts and fossil human material dating from about 200,000 years ago.

ARCHAEOLOGY The study of man's past; changes in his relation to the world around him, and the development of his cultures through time.

ARCHAEOMAGNETISM A dating method for fired clay hearths or kilns. It consists of using information on the magnetism of iron oxides which have been submitted to considerable heat in the past (i.e. their archaeomagnetism).

ARTEFACT OR ARTIFACT An object made by man.

ASSEMBLAGE Different artefacts found together in the same layer at one site. If the same assemblage is found at a number of sites, the tools may be described as an 'industry'.

ASSYRIA A name derived from the ancient city-state of Assur in Mesopotamia. Considerable development took place there during the second millenium BC.

AURIGNACIAN An important palaeolithic flint industry, dated to between about 40,000 and 25,000 years ago. It formed part of the material culture of peoples extending from France to the Balkans, and as far south-east as Iran and Afghanistan.

AUSTRALOPITHECUS An early 'hominid' of between about one million and five million years old, as yet the earliest fossil form known to be linked with the evolution of our own species. It was the first variety in the human line which made tools.

AXE Normally thought of as a cutting tool of stone or metal, but might at times have been purely for ritual purposes. The cutting edge is parallel to the haft.

AZILIAN A name denoting an early culture known to have centred on parts of southern France and northern Spain. Dated to the final part of the palaeolithic phase in Europe, it is especially known for its distinctive harpoons and painted pebbles.

AZTEC An early Central American tribe which entered the Valley of Mexico and was successful enough – before the Spanish conquest of the early 16th century – to rule a vast empire.

B

BANDKERAMIK A form of pottery with characteristic linear-spiral design, usually seen on gourd-shaped bowls and jars. It was made by people who were the first farmers in parts of eastern and central Europe (linear pottery culture, c 4500 BC).

BARROW Burial mounds, usually of a circular or long oval form. The burials are usually of the third or second millenium BC, but secondary burials are occasionally more recent. Far less common are mounds of Roman, Saxon, or Viking date.

BÂTON DE COMMANDEMENT An object, possibly for making leather thongs, found at certain European sites from the Aurignacian to the close of the upper palaeolithic. It is made of antler with a single rounded perforation, in some instances with considerable added decoration.

BATTLEAXE An axe designed specifically for warfare (or defence), particularly important as finds at European neolithic and Copper Age sites. The typical form, in stone or metal, has a rounded shaft hole, a cutting edge, and a blunted end. The same type of axe, in iron, proved of great value to the Vikings.

BEAKER An important form of pottery, usually found as a curved drinking vessel without a handle. It was widespread in Europe about 2000 BC, extending from Poland and southern Italy to the north of Scotland. The beakers are often associated in burials with metal daggers, arrows, and eventually with the battleaxe.

BELGAE Knowledge of this group of Celtic tribes is derived both from the writings of Caesar and from archaeological finds. The Belgae extended over parts of France and Belgium and finally moved into southern England about 100 BC. They were thus one of the important Iron Age communities of northern Europe, developing fortifications, towns, coinage, and the potter's wheel.

BLADE Part of the tool-kit of upper palaeolithic man. The long, thin stone flake was used directly as a working implement, or modified and used for specialized purposes.

BRECCIA Angular fragments of older rock held together by finer cementing material. Cave and rock-shelter deposits may be of this form, and some *Australopithecus* remains in South Africa occurred in breccias.

BRONZE AGE A term coined to emphasize the emergence of peoples who knew how to use this metal. It forms the intermediate general cultural level between the Stone Age and Iron Age. It is not easy to apply it outside Europe.

BURIN A sturdy stone tool with a point like a screwdriver, made by the intersection of two flake scars. It could be resharpened with a single blow. Many varieties were made by upper palaeolithic men for carving antler tools and engraving on cave walls.

C

CALENDARS A number of earlier societies were concerned with the measurement and recording of some division of time. Of most importance has been the cycle of a day, and of events related to the Sun and Moon. In Egypt, the civil calendar of 365 days was checked out against the movement of the star Sirius. In contrast to this is the ancient Calendar Round system of Central America, which combines the 260-day Sacred Calendar, of ritual importance, with the Solar Year Calendar of 365 days divided into 18-month divisions.

CARBON^{14} A radioactive isotope of considerable importance in dating precisely certain types of archaeological material of organic origin. Its value is in the time range between about 50,000 BC to 1500 AD, this being related to the fact that the C^{14} which is locked away in living matter at death decays at a known rate.

ÇATAL HÜYÜK A very early town site in central Turkey, extending back to at least 6150 BC. Farming was well developed; wheat and barley, legumes, sheep and cattle formed important parts of the diet. Ritual appears to have been important, and shrines with modelled bulls' heads were among the structures discovered.

CAVE ART A term usually associated with the palaeolithic art of Europe, and including paintings and engravings in rock shelters. Art of this kind is also seen in other parts of the world, and in the case of that in southern Africa may include depictions of Europeans – indicative of art work of only a few centuries ago.

CELTS An early group of people living in central and western Europe who developed the characteristic La Tène art style about 450 BC.

CHAMBERED TOMB A well-constructed burial vault, usually of large slabs of stone. Successive burials were usually placed in the individual chambers. Such tombs are found in various parts of the world and are of different dates.

CHELLEAN A term once widely used in relation to certain lower palaeolithic handaxe cultures. It is still used, for instance when referring to certain East African tools, but has been largely replaced by the term Acheulian.

CHICHÉN ITZA A major centre in Mexico where Mayan ceremonies were performed.

CHIMU A large and powerful empire in Peru. Developing after 1000 AD, it was conquered by the Inca people about 1470.

CHOPPER A large stone tool, usually of flint or chert, with a cutting edge produced by flaking from two sides.

CHOUKOUTIEN A site near Peking, China, which has yielded both Middle Pleistocene evidence of fossil man (*Homo erectus*), and important upper palaeolithic human remains. One of the earliest examples of fire use comes from this site.

CLACTONIAN A lower palaeolithic tool industry named after the industry found at Clacton-on-Sea in south-east England.

CODEX Any early Mexican manuscript prepared by Amerindians, and dating to pre-Spanish Conquest or the early colonial period. Early forms are on skin or bark paper.

CRANIAL CAPACITY The size of the brain as measured either by determining directly the internal volume of the brain box of the skull, or by the application of special formulae to certain external measurements of the skull.

CREMATION The ritual burning of the dead. The age of this practice varies widely in different parts of the world.

CRO-MAGNON A French rock shelter famous

for the early discovery of upper palaeolithic remains in association with human skeletons.

CUNEIFORM A characteristic form of writing originating in Mesopotamia and used between the third and first millennia BC.

D

DATING That aspect of archaeology concerned with chronology – that is, establishing times for the occurrence of peoples or particular objects. Relative dating – the ordering of events into a sequence – is being partly replaced by scientific absolute dating techniques such as Carbon^{14} and Potassium-argon methods.

DIFFUSION A recurring question in archaeology is whether a cultural trait, or characteristic, has been independently invented, or whether knowledge of it has been handed on and adopted from elsewhere – that is, diffused.

DOMESTICATION The control of plant or animal varieties by man, to the extent that noticeable changes can be seen to have occurred compared with original forms.

DORIANS Peoples who moved down through Greece after the decline of the Mycenaeans.

DYNASTY In archaeology, a succession of kings grouped under one name. The 30 dynastic groups of ancient Egypt are perhaps best known, but China, too, had long-lived dynasties. Regional dynasties also occurred in Mesopotamia.

E

EAST RUDOLF (East Turkana) Area in northern Kenya which has recently produced fossil hominid specimens and artefacts ranging in age from over 2 million to about 1 million years old.

ERTEBØLLE A mesolithic culture centred on the west Baltic area and characterized by shell middens and pottery.

ETHOLOGY The science concerned with the study of behaviour in animals.

ETRUSCANS A highly advanced society of north central Italy of the early first millennium BC. The Etruscans traded with both the Greeks and Phoenicians. Some Etruscan tombs are very rich in grave goods or art. The language is still a puzzle, and little has been translated.

EVOLUTION The emergence of one form from another. While in archaeology this can refer to changes in material objects, it is more useful with reference to biological change, especially of human evolution.

EXCAVATION The digging of deposits which have relevance to the full reconstruction of the early history of man. Ideally, these are stratified layers, with a minimum of disturbance. Because excavating is a destructive process, great care must always be taken in recording site details and the objects discovered as fully as possible.

F

FAYUM A depressed area with a lake, situated away from the west bank of the Nile in northern Egypt. Fossil primates are **common here, and include** *Propliopithecus, Aegyptopithecus, and Oligo*

pithecus. The early farmers also found this an attractive area.

FIBULA A long bone of the lower leg. It is also the name given to a form of brooch looking like a safety-pin, usually decorated and made of bronze.

FLINT Hard nodules of silica rock which can be found in and dug out of chalk deposits. Chert is very similar; both are easier than most rocks to work into stone implements by striking off flakes.

G

GERZEAN A culture named from the site of El Gerza in the Egyptian Fayum. The earlier predynastic Amratian culture gave rise to this new development about 3600 BC.

GLACIATION Concerned with the formation of ice-sheets or glaciers, and thus of fluctuations in the extent of cold climates in various parts of the world. A series of glaciations of varying extent and severity occurred during the Pleistocene period.

GRAVE GOODS In many earlier cultures objects have been buried with the dead, in the form of food offerings, utensils for the next world, or other objects.

GRAVETTIAN An upper palaeolithic tool industry, characterized particularly by the nature of the small pointed flint knife blades.

GRIMALDI A locality near Monaco which has produced upper palaeolithic human skeletons, one erroneously thought to be a Negro.

H

HADAR (Afar) Region in northern Ethiopia where recent excavations have recovered artefacts and parts of skeletons of *Australopithecus*.

HALLSTATT One of the most interesting of the Austrian Iron Age sites. From late Bronze Age times on, the area was a salt mining area. The cemetery of nearly 3000 graves has yielded large numbers of objects for study.

HANDAXE An all-purpose flint or chert tool, of oval or pear shape, and bifacially worked. Lower and middle palaeolithic cultures using these tools are found in Africa, Europe, and parts of Asia.

HARAPPA One of the capitals of the Indus civilization in the Punjab. Its walling, in parts of massive proportions, was of mudbrick.

HELLADIC The term referring specifically to the Greek Bronze Age.

HENGE Monumental structure of circular form, found only in Britain. In date henges appear to be restricted to late neolithic and early Bronze Age times. Stonehenge is the classic example.

HIEROGLYPHS A form of writing known especially from early Egypt. It appears to have evolved by about 3000 BC and continued through into early Christian times. The script was made up of pictograms – words in pictorial form – and phonograms representing the sounds of words. Some use the term also to refer to the pictograms evolved in Minoan Crete, and among the Hittites and the Maya.

HILLFORT A hilltop enclosure fortified with

ramparts and ditches. There is great variation in size. Hillforts were the first towns in central and northern Europe from the second millenium BC.

HITTITES A distinctive early group who moved into central Turkey and formed a state in about 1750 BC. This expanded into north Syria by 1450 BC, but decline set in by 1200 BC.

HOABINHIAN A mesolithic/neolithic culture in South-east Asia, named after the Vietnamese site of Hoabinh. It may well have been significant in the transfer of early farming knowledge farther south and eastwards.

HOLOCENE Also known as the Recent or Postglacial period; that time until the present which follows the Pleistocene.

HOMINIDAE The correct zoological term for the family or evolutionary group including man and his ancestors.

HOMO The name given to various species of hominid, most of which are internationally accepted as being sufficiently advanced to be considered as true man. The major species are *Homo erectus* and *Homo sapiens*, the former evolving perhaps over a million years ago and the latter certainly by Upper Pleistocene times. Other species names such as *Homo habilis* and *Homo neanderthalensis* are not acceptable to all human palaeontologists.

I

IBERIANS A group of people showing some cultural variation but with a common script and language. Their distribution was mainly in the more south-easterly parts of Spain; they flourished during the first millennium BC.

INCA A tribal dynasty established in the Cuzco region of the Peruvian Andes by AD 1200. Over the next 300 years, the empire of the Incas spread enormously as far as northern Ecuador and southwards to northern Chile. This was controlled by a central organization and ruled over by a divine king.

INDO-EUROPEAN The term usually refers to a group of languages originating in peoples in the Asiatic steppes, and spreading by second millennium times to Europe and the Near East.

INDUS CIVILIZATION The early advanced society of the Indus Valley. These early agriculturalists were well established by about 2300 BC and the twin miniature cities of Mohenjo-daro and Harappa had developed. Some degree of town planning had evolved, as illustrated by the positioning of houses and elaborate drainage.

INHUMATION The generally common practice in the past of burying the dead by one means or another.

INTERGLACIAL The period of warmer weather between two glacial periods during the Pleistocene, usually associated with changes in the vegetation and animal life.

IRON AGE A cultural phase of varying date depending upon the region of the world involved. In Europe it is prehistoric but in much of Asia historic, and was of European origin in America. This term is becoming less used.

J

JARMO An early farming site in northern Iraq. Carbonized wheat and barley were found, and possibly early domestic goats.

JERICHO Situated at the northern end of the Dead Sea, in the Jordan valley, the site has a long history of occupation beginning in the mesolithic (about 8000 BC). By neolithic times it had developed into a walled town, with mudbrick houses.

JŌMON A Japanese cultural phase lasting from about 7000 BC to late BC times. The evidence is from various coastal midden sites. Pottery was produced early, and has been drawn into the controversy on the possible trans-Pacific movement of people from eastern Asia into the west coast area of South America.

K

KHIROKITIA The site of an early farming (neolithic) community in Cyprus.

KNOSSOS The settlement in north Crete, known for its Minoan palace of 2000 BC. It was made famous by the excavations of Sir Arthur Evans between 1899 and 1935, which revealed a highly sophisticated and artistic civilization.

L

LASCAUX A cave in the Dordogne, France, famous for some of the finest upper palaeolithic paintings so far discovered.

LA TÈNE The term given to the second phase of the Iron Age in Europe. A distinctive abstract art style was evolved during this period.

LA VENTA On available evidence, this is the most important Olmec ceremonial centre so far found in Mexico.

LEPENSKI VIR An early settlement in Yugoslavia extending as far back as the seventh millennium BC with remarkable carved stones and trapezoidal houses. This was a fishing community on the banks of the Danube.

LINEAR A AND B Forms of script used by Mycenaeans in Crete and Greece.

LOS MILLARES A copper-using society of south-east Spain. A walled town and about a hundred passage graves have been excavated at the site.

LUNG SHAN Late neolithic people occupying part of the Yellow River area of China.

M

MAGDALENIAN Named after a site in the French Dordogne, this is the final cultural phase of the upper palaeolithic throughout much of western Europe. Its time range extends over about 5000 years, until about 10,000 BC. It was distinguished by polychrome cave paintings and carved antler tools.

MAGLEMOSIAN A stage of the northern European mesolithic, adapted to forest, lake, and river life. Its date range is from about 8000 to 5000 BC.

MAYA One of the important civilized Amerindian communities of Central America. By about 200 BC, the Mayans were constructing pyramids in the central lowlands of Guatemala. Although some of the population may have been well dispersed in the jungle, the great ceremonial centres must have acted as focal points for the whole community.

MESOLITHIC A general term indicating cultures of hunting and collecting economies in Europe adapting to changing environments after the end of the last glaciation about 10,000 BC. The beginnings of farming may have been occurring. Mesolithic cultures are characterized by microliths.

MICROLITH A small stone tool, often of geometric shape, and used especially during the mesolithic. These small flakes were used as barbs or points, perhaps especially on arrows.

MIDDEN A mound or layer of refuse, often including a considerable amount of food remains.

MINOANS The Bronze Age people of Crete.

MISSISSIPPI CULTURE Beginning in the river valley of that name, the culture became successful in the eastern United States generally, lasting from about AD 700 until European contact times.

MOHENJO-DARO One of the capital cities of the early Indus Valley civilization.

MOUNT CARMEL An area in Palestine whose caves have yielded the remains of fossil man which have proved critical to the proper understanding of human evolution during the early Upper Pleistocene.

MOUSTERIAN A stone industry, mainly of flint, comprising flake tools and at times handaxes. It is often, but not exclusively, associated with the Neanderthalers of Europe.

MUMMY Usually considered to be a body, either of man or another animal, which has been treated to some extent after death and wrapped in linen. In the case of the very early Egyptian bodies, and some from Peru, the preservation is simply the result of natural desiccation.

MUNGO Site in New South Wales, Australia, which has produced several cremated Pleistocene fossil human skeletons, one dating back to 30,000 years ago.

MYCENAEANS A term generally used to refer to the Late Bronze Age people of the more southern and eastern parts of Greece.

N

NATUFIAN The last phase of the mesolithic period in the Levant. As in other mesolithic peoples, microliths were important implements. Artefacts include reaping knives, probably for wild grain.

NEANDERTHAL A distinctive form of fossil man, mainly inhabiting Europe between about 80,000 and 35,000 BC.

NEOLITHIC A term defined in the 19th century to account for the last phase of stone-tool using before metal, and yet associated with an early farming economy.

NIAH A limestone area in Sarawak, Borneo, containing caves, one of which has yielded an advanced hominid skull thought to be about 40,000 years old – on C^{14} evidence.

NON NOK THA A site which has come to prominence in the last few years. Situated in northern Thailand, it shows evidence of animal domestication and the cultivation of rice by 3500 BC.

NUBIA The region south of Egypt, less developed but rich in gold, hard wood, and workable stone. It was also the trade gateway to ivory and ebony of central Africa.

O

OBSIDIAN A dark flint-like rock of glassy texture, which is of volcanic origin. It had most importance in the eastern Mediterranean and Central America, and was clearly of trade value.

OCHRE A pigment derived from red or yellow oxides of iron. It has been used as a colouring material since upper palaeolithic times.

OLDUVAI One of the most important areas to reveal hominid remains and stone tools in the world. Situated in Tanzania, East Africa, explorations of the gorge were made famous by Drs L. S. B. and M. Leakey. The stratified deposits probably span two million years.

OLMEC Although the Olmec people lived through into historic times, the great period of the Olmec civilization was early in the first millennium BC. It contributed significantly to the development of Mesoamerican culture as a whole.

OMO Area of southern Ethiopia where early hominids and artefacts have been discovered, as well as important Upper Pleistocene material of *Homo sapiens*.

P

PALAEOBOTANY Various botanical studies which have relevance to archaeological work. In particular, it includes pollen analyses, studies on carbonized wood and cereal grains, and the counting of ancient tree rings as an aid to dating.

PALAEOLITHIC This general cultural term covers the stone tool industries extending through the Pleistocene period. It can be roughly divided into lower, middle, and upper palaeolithic divisions.

PAPYRUS A reed which was valuable to some early communities in the eastern Mediterranean – especially the Egyptians. Used in various ways, including the making of light boats, it is best known as a common form of writing material, numerous fragments of papyrus having been preserved.

PEBBLE TOOL The earliest form of stone tool, made by striking a few flakes from a pebble to produce an irregular cutting edge. Its antiquity may go back to about three million years, and certainly it is associated with early hominids in Africa.

PEKIN MAN The name given to the regional variety of *Homo erectus* found in China.

PHARAOH Any of the Egyptian kings, figures who – to a greater or lesser degree – were regarded rather as gods.

PILTDOWN Village in Sussex where the remains of *Eoanthropus dawsonii* were discovered between 1910 and 1916. These were hailed as a primitive form of man but were later revealed to be the remains of a modern human skull and ape jaw 'doctored' to appear ancient.

PLEISTOCENE The geological period which covers the stages of hominid evolution at a more advanced level. Colder climates initiate this period, and its close is marked by the appearance of warmer Postglacial times.

POSTGLACIAL Any time after the Pleistocene period, beginning about 8000 BC.

POTASSIUM-ARGON DATING Like the radiocarbon method, this is an absolute dating technique. In this case the isotope K^{40} has a half life of 1300 million years. Its dating value is therefore much earlier than for C^{14}, extending back from about half a million years ago until before the beginning of the Pleistocene.

PREDYNASTIC A term used to cover the last phases of the prehistoric period in Egypt, dated to about 4000–3200 BC.

PREHISTORY That part of the story of man concerned with his life and culture prior to the time of written records.

PYRAMID A term usually used to refer to the large monumental tombs constructed by the ancient Egyptians. Step structures of a similar kind, but not of the same function, are also known from south-west Asia and Central America.

Q

QAFZA (Jebel Qafzeh or Kafzeh) Cave in Israel which has produced some of the oldest skeletons attributable to modern man, yet associated with a Mousterian industry.

QUATERNARY A geological era which includes the Pleistocene and more recent times.

QUERN One or two stones shaped for grinding corn. The earliest form was the concave saddle quern, to be followed by the rotary quern – where one stone rotates on another.

R

RADIOCARBON DATING See Carbon[14].

RAMAPITHECUS Possibly the oldest hominid, known from Miocene or Pliocene deposits in Africa, Europe, and Asia.

RHODESIAN MAN A variety of fossil man who may have been widely spread in more southern parts of Africa perhaps 100,000 years ago.

ROCK SHELTER A sheltering place formed by overhanging rock.

S

SCRAPER A form of stone implement with a restricted concave or convex working edge. It was probably used for scraping skins or working wood.

SCYTHIANS A group of nomadic people whose extent was in the area west of the river Volga and north of the Black Sea. Herodotus gave an account of them after a visit to the area in about 450 BC.

SHANG An early dynasty in China, dated to between about 1500 and 1027 BC. In terms of metal using, it was advanced Bronze Age, with well-developed towns. Writing was used, and thus Chinese history begins in this period.

SHERD A fragment of pottery vessel.

SOAN An early stone tool industry derived from the rough working of pebbles. Its distribution is centred in the Punjab and northern part of India.

SOIL ANALYSIS Useful in the reconstruction of past environments. Undisturbed soils beneath archaeological structures are especially valuable for analysis.

SOLUTREAN A skilled technique of producing long pressure-flaked bifacial spearpoints. Formerly thought to be an upper palaeolithic culture that originated in Hungary and moved westwards into France.

SPIRIT CAVE A site in north-west Thailand which has revealed possible evidence of early agriculture which may be 12,000 years old.

STAR CARR A British mesolithic site which has yielded a rich collection of bone, wood, and flint objects.

STEINHEIM A Middle Pleistocene site in Germany which is famous for the human skull found there.

STONE AGE A very general term meaning any culture in the past that used stone artefacts but not metal ones.

STRATIGRAPHY The study of successions or sequences of deposits at a site. The 'layers' or 'levels' may be cut into by subsequent deposits and thus caution is needed in interpretation.

STONEHENGE A large circular monument in Wiltshire, England, unique in its lintelled construction (which gave it its name of 'hanging stones') and in the dressing smooth of the surfaces of the stones by a four-stage process. Constructed between c 2100 BC and 1400 BC, possibly for astronomical observations.

SUMER An area in lower Mesopotamia where a civilized state of society was reached by about 3500 BC. Cities became large and complex, as at Ur.

SWANSCOMBE A site to the south of the river Thames where Clactonian and Acheulian artefacts, animal remains, and pollen have been recovered dating from an interglacial period. The remains of 'Swanscombe man' probably date from about 250,000 years ago.

T

TAUNG A limestone cave in South Africa where the first specimen of *Australopithecus* was discovered.

TEHUACAN VALLEY An area of Central America which has been especially well investigated archaeologically. Evidence of human habitation extends back from Spanish Conquest times to about 9000 BC.

TELL A mound which has developed as a result of the accumulation of debris (e.g. collapsed mudbrick) and refuse from long human habitation at one site.

TEOTIHUACAN One of the archaeological show pieces of Mexico, and originally a large urban centre. Ceremonial structures include the pyramids of the Sun and Moon.

THERMOLUMINESCENCE A dating technique which is still in the earlier stages of development. Perhaps the most important application of the method is to the dating of pottery by establishing the time when it was originally fired.

TIKAL A site in Guatemala which began as a small farming community but had evolved into a major Mayan ceremonial centre by the 1st century AD.

TOLTEC An early Mexican people of mixed origin, perhaps most significant for their introduction of metallurgy into the area.

TORRALBA A palaeolithic hunting site where early Acheulian tools were found in association with many bones, especially of elephants.

TRINIL Site in Java where the first *Homo erectus* remains were discovered by Eugene Dubois in 1891.

TUMULUS A structure covering a burial or burials, usually earth with varying use of stones.

TYPOLOGY The classification and comparison of different types of archaeological objects usually on the basis of shape differences.

U

UR A tell south of the river Euphrates. Occupation began about 4300 BC and the society was in decline by the first millennium BC. Spectacular finds were made in the Royal Cemetery of the early third millennium.

URNFIELD A cremation cemetery in which the majority of burned bones are placed in pots called funerary urns.

V

VÉRTESSZÖLLÖS A site near Budapest in Hungary which has produced human remains, perhaps belonging to the species *Homo erectus*.

VILLAFRANCHIAN The earliest phase of the Pleistocene period.

W

WINDMILL HILL A causewayed enclosure in Wiltshire, England, possibly built for periodic tribal gatherings and ancestral to the later henges. Famous for its wealth of pottery of the middle neolithic stage (3000–2500 BC) and for the evidence of stockbreeding and cereals grown at this time. The term 'Windmill Hill culture' is no longer used by archaeologists.

X

X-RAYS The use of X-rays in relation to archaeology is of two kinds. Radiography permits the internal structure of bone, pottery, and even metal to be seen. X-ray fluorescence is another technique, in this case of value in the chemical analysis of objects.

Y

YANG SHAO The neolithic culture of China, situated along part of the course of the Yellow River.

Z

ZAPOTEC An early Mexican people living mainly in the valley of Oaxaca. Monte Alban appears to have been the main political centre.

ZONES A term of particular value in environmental archaeology, concerned with changes in vegetational composition and climate through time – particularly in northern Europe. Pollen analysis established this sequence.

Index

Bold entries indicate a major mention

Italic entries indicate illustrations

ACKNOWLEDGEMENTS

Photographs: Page 1 Imitor; 2, 4 Michael Holford; 6 *top left* C. Stringer, *bottom left* Robert Harding Associates, *right* Michael Holford; 7 *left* Michael Holford, *top right* Zefa (UK) Ltd, *bottom right* R. Harvey; 8 Robert Harding Associates; 11 Foto-Call International; 13 Zoological Society, London; 16 Primate Research Laboratory, University of Wisconsin; 17 Camera & Pen International; 18 Norman Heitkotter; 21 Windsor Safari Park; 22 Sonia Halliday; 23 Zefa (UK) Ltd; 24 *top* Zefa (UK) Ltd, *bottom* Bulgarian Embassy; 25 *top* Nasa, *bottom* P. Morris; 26 Michael Holford; 27 *top* Institute of Geological Sciences, London, *bottom* British Museum; 28 Oriental Institute, University of Chicago; 29 *bottom* Robert Harding Associates; 30 N. Seeley; 32 Institute of Geological Sciences, London; 33 Tom Myers/TSA: NHPA; 34 British Museum; 36 Finnish Embassy; 37 The Royal Free Hospital; 43 Bruce Coleman; 46, 47, 48, 50, 51 P. Andrews; 54, 55 C. Stringer; 56 P. Andrews; 57, 58 C. Stringer; 59 *top* P. Andrews, *centre, right, and bottom* C. Stringer; 63 Imitor; 64 P. Andrews; 65 C. Stringer; 66 L. Angel; 67 C. Stringer; 68 Imitor; 69 *top* Imitor, *bottom* C. Stringer; 70, 72 C. Stringer; 73 *top* Zefa (UK) Ltd, *bottom* C. Mountford; 74 *top left and right* R. Harvey, *bottom centre* C. Stringer; 76 *top* Zefa (UK) Ltd, *bottom* C. Stringer; 77 *top* C. Stringer, *bottom* J. Picton; 78 P. Tobias; 79 Robert Harding Associates; 80, 81 Jean Vertut; 84 Novosti; 86 *top* Jean Vertut, *bottom* Moravian Museum, Czechoslovakia; 87 *top* Caisse Nationale des Monuments Historiques, Paris, *bottom* Jean Vertut; 88, 89 Jean Vertut; 90 Michael Holford; 91 *top* Giraudon, *bottom left* National Museum, Copenhagen, *bottom right* Photoresources; 95 *top* R. Place; 96 Zefa (UK) Ltd; 97 *left* J. Powell, *right* Oriental Institute, University of Chicago; 98 *top* P. Clayton, *bottom* Sonia Halliday; 99 *right* J. Mellaart, *bottom* A. M. T. Moore; 102 *top* Devizes Museum, *bottom* Michael Holford; 103 *centre* Robert Harding Associates, *top* Institute of Geological Sciences, London; 105 National Museum, Denmark; 106 *top* British Museum, *bottom* Ashmolean Museum, Oxford; 107 *top* J. Powell, *centre* British Museum, *bottom* Robert Harding Associates; 108 Naturhistorisches Museum, Austria, *bottom* National Museum, Wales; 109 Robert Harding Associates; 110 *top* Michael Holford, *centre* Peter Clayton, *bottom* Ashmolean Museum, Oxford; 111 *top* Michael Holford, *centre left* Peter Clayton, *centre right* Michael Holford, *bottom left* British Museum, *bottom centre* British Museum; 112 *top* Peter Clayton, *centre* Michael Holford, *bottom* Robert Harding Associates; 113 *left* National Museum, Karachi; *right* Robert Harding Associates; 114 *bottom right* Picturepoint; 115 *top* Sonia Halliday, *bottom* Michael Holford; 116 Lord William Taylour; 117 *top* Sonia Halliday; 118 *top* Department of the Environment, *bottom* National Museum, Copenhagen; 119 *top* Schleswig-Holstein Museum, Germany, *centre and bottom* Bulgarian Embassy; 120 *top* Naturhistorisches Museum, Austria, *bottom* British Museum; 121 *top* British Museum, *bottom* National Museum, Copenhagen; 122 *top* Robert Harding Associates, *centre* I. Glover, *bottom* Tokyo National Museum; 124 I. Glover; 125 *top* Tom Haydon, *centre* British Museum, *bottom* Tom Haydon; 126 *top* British Museum, *centre* Society for Anglo Chinese Understanding, *bottom left* Museum of Far Eastern Antiquities, Stockholm, *centre* Society for Anglo Chinese Understanding, *bottom* Zefa (UK) Ltd; 128 *top left and centre* Don Bayard, *bottom* Robert Harding Associates; 132 Satour; 133 Michael Holford; 134 *top right* British Museum, *bottom left* British Museum, *bottom right* C. R. Warn; 135 *top* Peter Clayton, *centre* British Museum, *bottom* Peter Clayton; 137 British Museum; 138 *top right* Zefa (UK) Ltd, *centre* British Museum, *bottom* Egyptian Tourist Office; 139 W. Bray; 141 *top* British Museum, *bottom* Michael Holford; 143 British Museum; 144 *top* Robert Harding Associates, *bottom* E. Carter; 145 British Museum; 146 *left* Michael Holford, *centre* W. Bray; 147 *top* Zefa (UK) Ltd, *centre* British Museum, *bottom right* Michael Holford, 148 W. Bray; 149 *top* Michael Holford, *bottom* Robert Harding Associates; 150 Michael Holford; 151 *top* Michael Holford, *bottom left* W. Bray, *bottom right* E. Carter.

Picture Research: Penny Warn and Jackie Newton